Finding Water
Second Edition

Finding Water
Second Edition

A Guide to the Construction
and Maintenance
of Private Water Supplies

RICK BRASSINGTON

JOHN WILEY & SONS
Chichester · New York · Brisbane · Toronto · Singapore

First edition © 1983 Rick Brassington. Published 1983 by Pelham Books, Ltd

Second edition © 1995 by John Wiley & Sons Ltd,
              Baffins Lane, Chichester,
              West Sussex PO19 1UD, England
              Telephone National   01243 779777
                      International (+44) 1243 779777

Other Wiley Editorial Offices

John Wiley & Sons, Inc., 605 Third Avenue,
New York, NY 10158-0012, USA

Jacaranda Wiley Ltd, 33 Park Road, Milton,
Queensland 4064, Australia

John Wiley & Sons (Canada) Ltd, 22 Worcester Road,
Rexdale, Ontario M9W 1L1, Canada

John Wiley & Sons (SEA) Pte Ltd, 37 Jalan Pemimpin #05-04,
Block B, Union Industrial Building, Singapore 2057

Library of Congress Cataloging-in-Publication Data
Brassington, Rick.
    Finding water : a guide to the construction and maintenance of
  private water supplies / Rick Brassington. — 2nd ed.
      p. cm.
    First published by Pelham Books, Ltd., c 1983.
    Includes bibliographical references and index.
    ISBN 0-471-95711-9
    1. Wells. I. Title.
  TD405.B72 1995
  628.1—dc20
                                      94-49357
                                      CIP

British Library Cataloguing in Publication Data
A catalogue record for this book is available from the British Library

ISBN 0-471-95711-9

Typeset in 11/13 pt Palatino from author's disk by
Mathematical Composition Setters Ltd, Salisbury, Wiltshire, SP3 4UF
Printed and bound in Great Britain by
Biddles Ltd, Guildford and King's Lynn

Contents

Preface   vii

Acknowledgements   ix

1   Introduction   1

2   The Hydrological Cycle   6

3   Choosing the Most Suitable Source   21

4   How Much Do I Need?   62

5   Building a New Source   72

6   Water Quality and Treatment   115

7   The Rest of the System   139

8   Maintaining your System   186

9   Water Rights   195

10 Problems with External Causes   222

Appendix 1   Conversion Factors   244

Appendix 2   Safe Working Practices   246

Reading List   251

Glossary of Technical Terms   253

Index   265

Preface

During the summer of 1981, I was asked to visit a farmer in the Yorkshire Dales who had a problem with his water supply. His farm is supplied with water from a spring which had never been known to fail during either his time or that of his father, but had now dried up. He blamed the effect on a new borehole nearby, but as soon as I saw his source I realised that his problem was simply a lack of maintenance. There was plenty of water but it was no longer flowing into his catchpit; instead it had turned the surrounding area into a quagmire. If he had kept the inlet pipe clear of silt there would not have been a problem. The first edition of this book resulted from me trying to help that farmer.

In a newspaper article printed not so long ago, a householder who uses a private water supply was quoted as saying that "The water is the colour of tea but tastes fine so long as no sheep have fallen into the burn (stream) recently!" There is absolutely no reason why anyone who uses streams, springs, rivers or wells for their water supply should have such a poor standard of living. Perhaps that press comment is not typical of most private water supplies but it underlines the type of problem with which some people are faced. Water is a very important commodity in all our lives, but one which is too easily taken for granted. As far as private water supplies go, this has often meant little or no maintenance. When problems appear they may cause great difficulty in trying to resolve them and perhaps an acceptance of an unpleasant taste or a supply insufficient to fill a bath.

The main purpose of the book is to help people who rely on their own source of water to have a pure and wholesome supply, and one which is reliable and sufficient to meet their needs. The first edition of this book was specifically aimed at British farmers and country dwellers. This new and extensively revised edition contains much more detailed information and has been written for an extended readership and should provide practical help for people living in most countries round the world. It covers all aspects of building and maintaining small water supplies for individual properties, farms and small business users and should also be useful to people looking for their own water supply to become independent of the public mains.

Rick Brassington
Warrington
November 1994

Acknowledgements

A number of people have helped me during the preparation of this book and in obtaining information. I am indebted to Penny Bland and Rita Clements for helping transfer the text to a word processor, David Brassington for solving my numerous word-processing problems and Fred Brassington for critically reading the manuscript. My wife Sandra, has given me her support and helped with typing, proof-reading and in many other ways. My thanks to her acknowledge that she continues to patiently tolerate the domestic disruption that is associated with writing a book.

A number of people have provided me with information and I am grateful to Sharron Andrews of DANDO Drilling International Limited, Littlehampton, West Sussex; Ken Beardsell of Hepworth Industrial Plastics Limited, Sheffield; Tony Beer of Southern Science Limited, Worthing; Neal Burton of Intermediate Technology Publications Limited; Chris Dodds of Dales Water Services, Melmerby, Ripon; John Hills of British Water, London; Professor Ken Howard of the University of Toronto, Canada; Chris Stevenson, H_2O Waste-Tec (Mono Pumps Limited), Stockport, Cheshire; Dr T. Kubo of the Japanese Sewage Works Association, Tokyo, Japan; Frank Law of the Institute of Hydrology, Wallingford; George Reeves of the University of Newcastle upon Tyne; Kenji Sambongi of the National Diet Library, Tokyo, Japan; Alan Simcock, Head of Water Resources and Marine Division, Water Directorate,

Department of the Environment, London; Rodney Sleigh of John Blake Limited, Accrington; Dr John Suter of Carl Bro Haiste, Leeds; Dr P. Taylor of the Institute of Water and Sanitation Development, University of Zimbabwe; Mark Whettall of Durapipe, Glynwed Plastics Limited, Cannock; Duncan Wormald of Wavrin Industrial Products Limited, Durham; and Colin Wright of the Water Resources and Marine Division, Water Directorate, London.

A number of figures and tables have been adopted from or inspired by the following sources, to whom grateful acknowledgement is made: Figure 3.1: P. Beaumont, Qanats on the Varamin Plain, Iran. *Transactions of the Institute of British Geographers*, No 45, pp. 169–179 (1968); Figure 5.15: Photographs provided by DANDO Drilling International Ltd; Figure 5.17: Photograph provided by Dales Water Services Limited; Figure 5.18: L. Clark, *The Field Guide to Water Well and Boreholes*, John Wiley & Sons (1988); Table 6.1: N. Gray, *Drinking Water Quality — Problems and Solutions*, John Wiley & Sons (1994); Table 6.2: Based on Canadian Water Quality Guidelines (1987), Canadian Council of Resource and Environment Ministers, reproduced by permission of the Minister of Supply and Services 1995; Figures 7.7, 7.8 and 7.9: Photographs and drawings provided by John Blake (Allspeed Limited); Figure 7.11: redrawn from *The Village Carpenter* by Walter Rose, Cambridge University Press (1937, 1973 and 1987); Figure 7.15: Material supplied by H_2O Waste-Tec (Mono Pumps Limited); Figure 7.20: Information supplied by Durapipe, Glynwed Plastics Limited, Hepworth Industrial Plastics Limited and Wavin Industrial Products Limited; Table 7.1: T. D. Jordan, *A Handbook of Gravity-Flow Water Systems*, Intermediate Technology Publications (1980), by permission of UNICEF Kathmandu; Figures 9.2, 9.3 and 9.4 and Table 9.1: K. R. Wright (ed.), *Water Rights of the Fifty States and Territories*, American Water Works Association (1990); Figure 10.7: Drawing provided by WaterAid, London.

1
Introduction

Everyone needs water! It is essential to keep people alive and is just as vital for plants and animals. In fact, a water supply is needed for houses, farms and just about everywhere people live or work. The better the water supply, the better the quality of life. Just think of how many different jobs water does in most homes then try to imagine the headaches which would be caused by only having a trickle of poor quality water to use. Water is not only needed for drinking but for cooking, and washing pots, pans, people and clothes. It is used for sanitation, cleaning the house, washing the car, watering the lawn and a thousand other jobs. On the farm, ranch and smallholding, water is needed for watering the animals, cleaning the dairy and irrigating crops. In business and industry it also plays a hundred other vital roles.

Despite its importance, most of us give very little thought to our water supply beyond turning on the tap and grumbling about the size of the water bills. So long as the water which comes out of the tap can be drunk safely and there is enough for us to use, we give no further thought to it. This is understandable if you happen to be one of the majority of people in most western countries who have a water supply from the public mains. In virtually every country, however, there are a large number of people who do not have a mains supply and rely on their own source of water. For them the situation is very different.

It is difficult to obtain official statistics on the number of people who do not have a public or municipal water supply. On a world-wide scale it was estimated that around 1 500 000 people in develop-ing countries did not have an adequate water supply before the United Nations Water Decade of the 1980s, with an even larger number not having adequate sanitation services. The efforts of the Water Decade made significant progress in rectifying this situation although it will require many more years before the world's population all have safe and reliable water supplies.

On the face of it, few would think that there are many people in Britain who do not enjoy a water supply from the public mains. Indeed, the vast majority of households in this country have a piped water supply. Government figures show that 99% of the population in England, Scotland and Wales and 90% of the people living in Northern Ireland get their water in this way. This is an impressive record and the British water industry boasts that it provides the best service in the world.

These statistics can be used to work out how many people do not have a mains water supply. They show that there are three-quarters of a million people living in the United Kingdom who rely on their own wells, springs or other sources for their water supply. As may be expected, most of them live in rural areas, although not neces-sarily remote from towns or cities. In some hilly areas, such as the Pennines, there may be a number of houses just beyond the fringe of a town which are simply too high up to be supplied from the public system. In other places, the density of the population may be too small for the economic installation, maintenance and operation of a public system.

These problems are not new. Properties in these areas usually have their own water sources, often dating back well over a century and sometimes even longer. These old sources are usually springs, although there are also a large number of shallow dug wells that may have been built with the house. In contrast, more modern sources are usually boreholes, that is deep, small-diameter wells which have been constructed using a special drilling rig. These boreholes are usually lined with steel pipe which is between 100 mm and 400 mm in diameter and extend 25 m or more in depth. The reason for this modern emphasis on boreholes is that they are easily constructed by specialist contractors who are to be found in almost all areas. Furthermore, boreholes which have been properly constructed are unlikely to be affected by pollution and as such may attract grants from central government more easily than other sources. There are no reasons, however, why supplies to a new property or replacements for an old system should have to be boreholes.

There are a significant number of people in various countries across Europe who are not connected to a public water supply system. If we take a few examples based on official statistics we see that there are 3.25 million people in Austria, 200 000 in Belgium, 400 000 in Denmark, 940 000 in Finland, 640 000 in Hungary, 5.25 million in Italy, 1.6 million in Spain, 250 000 in Sweden, 64 000 in Switzerland and 1.25 million in the western part of Germany. From this review we can see that the use of private water supplies in rural areas is widespread and involves large numbers of people in all countries.

A significant proportion of the population in the United States of America also abstracts its own water, with the majority being taken from wells and boreholes. Besides private water supplies feeding individual properties there are also a large number (140 000) of small public supply systems which provide water for 25 or fewer people. In Canada the majority of the people living in rural areas also rely on their own water source. The same picture applies to Australia and New Zealand where the urban population is served by a municipal water supply and those who live in rural areas mainly fend for themselves. In Zimbabwe, there is a similar story with the majority of the rural population obtaining their water from primary supplies, such as a borehole fitted with a hand-pump, and in urban areas there are usually municipal piped water supplies.

HOW MANY SUPPLIES?

In trying to work out how many private water supplies are in use it is necessary to do a bit more detective work as the official statistics are again vague.

We have assessed the number of people who drink water from private supplies but the total number of private sources is quite a different figure. We have already assumed that in the United Kingdom there are some three-quarters of a million dwellings which rely on private water supplies. Besides these, there are a very large number of abstractions made by industry and farmers to supply their own needs. There are no official statistics of the total number of private sources and it is difficult to estimate accurately the number which are in current use.

The majority of abstractions are made without control from central or local government, but in England and Wales there is a licensing system which controls the larger abstractions. Official records of the number of licences which have been issued can be used to estimate the total number of private sources which are in current use. In England and Wales there are some 7000 licensed water abstractions made by industry. These are split almost equally between surface water (i.e. rivers, streams and the like) and groundwater (i.e. water that has been pumped from wells or boreholes). There are more than 32 000 further licensed sources used for agriculture, but these figures are misleading. The majority of these abstractions are from groundwater sources, as surface water sources which are in use for agriculture do not require licences. It is quite likely that the total number of agricultural abstractions in England and Wales exceeds 250 000. If we use these calculations for England and Wales, we can estimate that the total number of private water supplies in the United Kingdom is likely to exceed 750 000 and may even be more than one million.

It is difficult to come up with estimates of the total number of sources in use in various countries around the world. In general, however, there are very large numbers, counted in millions in most countries when abstractions made by industry and farmers are included.

EMPLOYING AN EXPERT

In several parts of the book it is made clear when it is necessary for an expert to be employed. It is important to select the appropriate expert, but this is not always easy, so here is some advice on how to choose the right one for the job.

One of the most satisfactory ways of deciding which contractor or consultant to employ is on the recommendation of previous clients. It is important, however, to make sure that the advice and service which he/she gave to these people is the same as you require.

If you are seeking advice from a consultant, ask to look at his or her qualifications and find out if they are a member of the appropriate professional institution. As professionals, consultants will be bound by the institution's code of professional conduct which means, among other things, that if their experience is not adequate to solve your problem they will tell you and perhaps suggest someone else to do the job. These considerations are important whether you are looking for an engineer, a chemist, a geologist or a lawyer. If you need help in finding a list of names, write to the professional institution most closely involved.

TECHNICAL TERMS

This book is about a technical subject but is written for the non-specialist. Every effort has been made to avoid the use of jargon and keep technical terms to a minimum. A glossary of technical terms is included at the back of the book and each term is highlighted in italics and explained each time it is used for the first time in the text. Perhaps the most confusing terms are *well* and *borehole*. Both are holes in the ground especially made for abstracting water. The essential difference lies in the method of construction. Wells have a wide diameter, are shallow and have often been dug by hand, whereas boreholes have a small diameter, are usually deep and are constructed using specialist machines. However, boreholes are a variety of water well and, except where it is made clear, comments made in this book about wells apply equally to boreholes and vice versa.

2
The Hydrological Cycle

Before you can start to look for a new source of water or to try and solve problems with an existing source, you really ought to appreciate both where the water comes from and where it is going. I have put it in this way to underline the fact that all water, wherever it is, forms a vast worldwide system known as the *hydrological cycle*. Figure 2.1 illustrates the various parts of this cycle, which both starts and ends with the oceans.

Energy from the sun causes water to *evaporate* (i.e. turn into vapour) from the surface of the world's oceans to form large cloud masses. These clouds are moved round the world by the global wind system and, when conditions are right, the water falls back to the surface again as rain, snow or hail. It is a natural process of fundamental importance to all life on the land.

Once the water has fallen onto the earth, it collects to form streams and rivers which eventually flow into the sea and allow the process to start all over again. This order of events does actually take place, but frequently the cycle is very complicated and has many loops, whorls and U-turns in it. For example, some of the water can bypass part of the system by falling as rain directly into the sea, rivers or lakes. Some of the rain which has fallen onto the ground, on the other hand, may be quickly evaporated back to the atmosphere, thereby cutting out another section of the cycle.

There are wide variations in the speed at which water moves round the cycle. Large volumes of water are trapped for very long

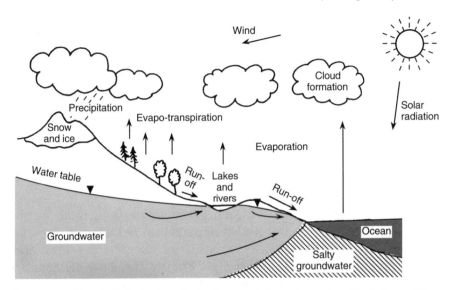

Figure 2.1 The hydrological cycle is illustrated here in a simplified form. Water evaporates from the oceans to form clouds and will eventually fall as rain, snow or hail. A significant part is lost by evaporation or taken up by plants as transpiration. In high latitudes and in mountain areas part of the water is held as ice and snow. The remainder either collects to form lakes, streams and rivers before flowing back to the sea, or soaks into the ground to become groundwater. The groundwater flows more slowly either to discharge into streams and rivers or directly into the sea. In coastal areas fresh groundwater is replaced by sea-water

periods in the polar ice caps and glaciers in high mountain ranges. The ice cap in Greenland, for example, is about 1400 m thick and the ice at the bottom has been shown to be about 150 000 years old. This gives you an idea of how water can be literally frozen out of the active part of the hydrological cycle for very long periods. At the other end of the spectrum, water which falls in a tropical rainstorm can easily be evaporated back into the atmosphere in a space of a few hours. Even in individual countries such as Britain, large contrasts exist between water which spends all winter on a frozen Scottish mountain and that which evaporates from roadside puddles as they quickly dry up after a summer thunderstorm.

At first glance it may seem that how quickly or slowly water moves through parts of the hydrological cycle is not really relevant to sorting out your new water supply but, because all water forms part of a continuous cycle, if you take water out of one part, you are likely to cause a "knock-on" effect and reduce the amount further down the line. The speed with which water is moving through that part of the cycle will control how quickly these effects come about. An obvious example is when someone pumps water from a stream which will reduce the flow downstream from the point where the water is taken out. A less obvious case may be where pumping from a well causes a reduction in the flow from springs which lie several kilometres away.

CLIMATE

Just what sort of weather we each get at home is dictated by the atmospheric circulation but it also depends on *latitude* (i.e. how far south or north of the equator you are), the *elevation* (i.e. how high above sea level you are), and the geography (i.e. how far you are from the sea and your position in relation to the continents). The western and eastern continental seaboards have different types of climate both from each other and from the mid-continental areas. If you live in the UK or Vancouver, for example, you will have a somewhat wetter and warmer climate than Japan or the eastern seaboard of the USA. If you live a long way from the sea, such as the American Mid-West or central Europe, the winters will be much colder and a significant proportion of the precipitation will fall as snow. In the Tropics, rainfall is much more seasonal and occurs during the monsoon period. All these factors influence and change the detail of how the hydrological cycle works in your part of the world but the main elements of precipitation, evaporation, surface run-off and groundwater recharge remain the same. Clearly climate is important, and the best source of water and the design of the means of abstraction is likely to vary from one location to another, but the principles explained in this book will help you assess the availability of water supplies wherever you may live.

PRECIPITATION

All water supplies start off as rainfall — or more accurately as *precipitation*. This wider term includes snow, sleet, hail, fog and dew, which all add to the quantity of water available for supplies. The proportion of total precipitation made up of any of these elements depends on local climatic conditions and will vary throughout the year and from place to place. In Britain, as most people are aware, the majority of precipitation falls as rain. The distribution of this rainfall, however, is very varied indeed, both in its geographical dispersal and from month to month or year to year.

In most countries, rainfall records are collected by government agencies. In the United Kingdom for example, the Meteorological Office collects information obtained from a network of more than 6500 rain-gauges. The majority of these rain-gauges are read at the same time each morning by an army of volunteers with the gauges sited in their gardens, school grounds or similar places. Additionally, there are a number of recording rain-gauges which register how long it takes for 2 mm of rain to fall so that more detailed information than daily totals can be used. All these readings are used as the basis for all rainfall statistics. The Meteorological Office publishes this information annually in a booklet and also in the form of maps which show the distribution of long-term average rainfall over the country. A more modern method of rainfall measurement uses a radar system to detect rain while it is still falling to the ground, but this system is unlikely to replace the rain-gauges for many years yet. Average annual rainfall ranges from less than 500 mm in parts of East Anglia to over 2500 mm in the Scottish and Welsh mountains and the Lake District. Even greater variations can be expected across continents such as North America where average precipitation varies from over 2500 mm in the Rocky Mountains to less than 250 mm in the desert areas in the western states. Precipitation also increases eastwards from around 500 mm in the High Plains where droughts are relatively common to over 1250 mm along the East coast.

Although water supplies ultimately depend on the quantity of rainfall in a locality, it may not be necessary to know exactly how much rain falls. The reliability of a water supply may be better judged by looking at things such as stream or spring flow records rather than rainfall statistics. If you want to know the quantity of average rainfall in your area, however, you are quite likely to be able to find out from your local reference library, government offices, water supply organisation or from the science teachers in your local high school or college.

Fog and dew are two forms of precipitation which are often missed out of discussions about the hydrological cycle. Both are related to the *humidity* of the air, i.e. how much moisture it can hold.

Humidity depends largely on temperature, with warm air being able to hold more water vapour than cold air. As a result, when saturated air cools, the amount of water vapour it can retain decreases and condensation takes place. The temperature when condensation starts is known as the *dew point*. As air cools to this temperature fog starts to form. Dew forms when air which is in contact with a cold surface such as the ground is cooled to its dew point; consequently dew may form without there being any fog. Fog is often associated with coastal areas because of the potential temperature contrasts between the sea and the land. It is also frequently found in hilly areas, caused by air cooling as it rises to flow over the hills.

EVAPORATION

Almost as soon as it has fallen as rain, some of the water will quickly return to the atmosphere by *evaporation*, that is it will turn back into vapour. Water will evaporate from a land surface whether or not it has a vegetation cover, and will also evaporate from the leaves of trees, roofs, roads, open water and even flowing streams. The rate of evaporation will depend upon the type of surface and whether it is in direct sunshine or in the shade. In many parts of England evaporation from open water may typically be the equivalent of 45 cm of rain per year.

The amount of evaporation will vary quite considerably from one place to another as it depends on several meteorological (or weather) factors. The energy used to evaporate the water comes from the sun and it follows, therefore, that on cloudy days evaporation will be much less than on sunny days. Air temperature and the temperature of the ground are also important as evaporation is greater at high temperatures than at cool ones. Wind plays an important part in evaporation processes and on windy days the rate of evaporation can increase quite significantly. Another meteorological factor which is important is relative humidity, i.e. the amount of water vapour in the air. As the air's humidity increases, its ability to absorb more water vapour decreases and the rate of evaporation slows down. For a given set of temperature and humidity there is clearly a limit to the potential amount of evaporation which can take place. However, there is a further limit to the amount of evaporation due to the amount of water actually present. As a result, in countries such as those in the Middle East where the potential evaporation during the year is much greater than the total rainfall you might expect there to be no water left. However, the rate of evaporation is significantly reduced during the period of rainy weather, producing enough water to form temporary streams and also leave some water to soak into the ground to become groundwater.

TRANSPIRATION

All kinds of growing plants need water to sustain life, although different species may have very different water requirements. Only a small proportion of the water needed by a plant is retained in the plant structure. Water is drawn into the plant through the roots and up through the stem or trunk to the leaves from where it is transpired into the atmosphere. *Transpiration*, in fact, is merely evaporation from the leaves. This action generates a "pulling" force which draws water up through the plant enabling water to be taken in through the roots. As this water travels up the plant, water and dissolved minerals are absorbed into the plant structure. To give you an idea of the quantities of water involved, a large deciduous tree, such as an oak or chestnut, growing in the open can transpire up to 600 litres of water vapour through its leaves on a single hot summer's day.

EVAPO-TRANSPIRATION

In field conditions where the ground is covered by vegetation, it is practically impossible to tell the difference between evaporation and transpiration. The two processes are commonly linked together and referred to as *evapo-transpiration*. The quantity of water which is potentially available for your water supply is the rainfall minus the evapo-transpirational losses. This is usually called the *effective rainfall*. If you are estimating the yield of a stream catchment entirely from rainfall, you will need to take into account losses caused by evapo-transpiration. These considerations will be particularly important when estimating the amount of water which percolates into the ground to replenish groundwater resources. In most instances you need not be too concerned as either you will be able to estimate the yield of a spring or stream by direct measurement or, if the problem is complicated, you will have called in an expert to do the calculations for you.

SURFACE WATER

This category includes all water which is flowing in rivers and streams or which is contained in lakes and ponds. All water which flows in a river comes from rainfall but it can take a variety of routes on the way. Rain may fall directly into surface water but some falls onto the ground and flows across the surface into the nearest channel. Water in this category is termed *surface run-off*. Part of the flow of a river is made up of groundwater which either flows from springs or enters the river directly through the bed as seepage. This groundwater contribution to river flow is termed *base flow* and, during periods of dry weather, can constitute the only natural water in a river. The other major part of river flow is termed *interflow*. It is made up of that water which percolates through the ground before reaching the river but has not reached the water table to become part of the groundwater; an example of interflow could be water from field drains. The term includes water which has moved out of a river and percolated into the banks as the level rises, only to flow back into the river as the level falls once more. The three parts of river flow — surface run-off, interflow and base flow — respond differently to periods of rainfall. Surface water reaches the river the

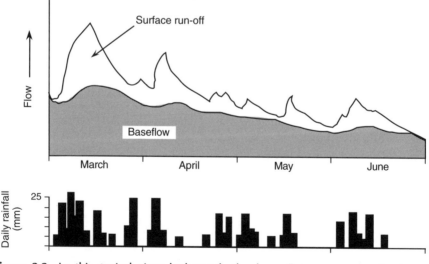

Figure 2.2 In this typical river hydrograph the base flow component is shown separately. Daily rainfall is indicated in the lower part of the diagram and by comparing the two you can see how river flow responds to rainfall

most rapidly and causes the initial increase in flow, interflow is next to respond to rainfall and base flow increases sometime after the period of rain. The relationship between rainfall and the components of river flow is illustrated in Figure 2.2.

GEOLOGICAL CONSIDERATIONS

The proportion of groundwater in a stream or river depends on the geology of their drainage basin or catchment area. Catchments containing a high proportion of porous rocks which allow water to flow through them easily, generally have rivers with hydrographs

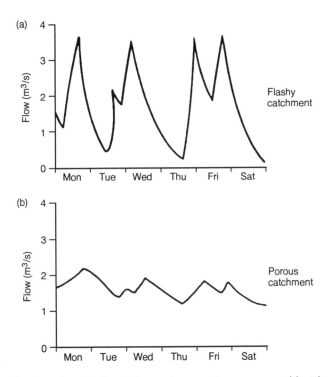

Figure 2.3 The river flows in the flashy catchment (a) are very variable. There are high peaks caused by rapid surface run-off and the river is liable to dry up completely in dry weather as there is little groundwater contribution to flows. The river in the porous catchment (b), however, is not likely to have such high peaks but will maintain a steady flow during dry weather. Here a higher proportion of the rainfall percolates into the ground, providing a large groundwater reserve which is available to maintain flows during dry weather

that are typified by high base flows and relatively low flood flows. This is caused by most of the rain percolating into the ground and flowing downwards to the water table to replenish the groundwater. This water will then flow through the rock to emerge again at the surface as springs or seepage directly into rivers or the sea. Rivers with catchments which consist mainly of impervious rocks will be flashy and have high flood flows but very low dry weather flows. These rivers and streams may even dry up completely in prolonged periods without rain. Figure 2.3 shows the difference in the hydrographs for flashy and porous catchments.

GROUNDWATER

Once rainwater has percolated through the soil down to the *water table* it becomes *groundwater*. In other words, groundwater is that water which flows through saturated rock under the influence of a *hydraulic gradient* which is, in fact, the slope of the water table. The water table is the upper surface of fully saturated rock. The water will stand in your well or borehole at the level of the water table. As we have already seen, groundwater is the source of spring water and it also supplies water which is pumped from wells and boreholes. The vast majority of private water supplies utilise springs, wells or boreholes so groundwater is the most important part of the hydrological cycle for us to consider in some detail.

Rock which contains groundwater and allows the water to flow through it is termed an *aquifer* and the capacity of a rock to transmit water is called the *permeability*. The way water flows through an aquifer is controlled by geological factors. Rocks such as sand and gravel have minute *pore spaces* between individual grains and water is able to flow through these spaces. Pore spaces can also be found in rocks such as sandstones, although the total volume of the pores is likely to be less than in an equivalent volume of sand because solid rocks are held together by cementing minerals which take up part of this pore space. In solid rocks there are frequently cracks, joints and other fissures along which water can flow. These fissure systems can play an important part in the total ability of the rock to transmit water. Indeed in some crystalline rocks, such as limestones and granites, groundwater can only flow through these fissure systems and is unable to move through the body of the rock. The flow of water out of all solid rock aquifers into a well or borehole is likely to be dominated by fissure flow and the presence of fissure systems is extremely important to well and borehole design.

Some types of rocks do not transmit groundwater at all or only allow small quantities to flow through very slowly. These rocks are termed *aquicludes* and *aquitards* respectively. Although they do not transmit much water, they both play a major role in controlling the movement of water through aquifers. Very few natural materials are completely uniform and most aquifers contain aquiclude and aquitard material which influence water movement through the rock. Figure 2.4 shows how a *perched water table* can exist above the main water table in an aquifer and the presence of an aquiclude, such as clay or unfissured bedrock, can give rise to springs. The reliability of a spring or a well as a source of water will depend on the extent of the groundwater body which it taps. If water supplies are taken from a spring which drains a perched water table or a well which taps one, they are unlikely to be reliable and may fail during

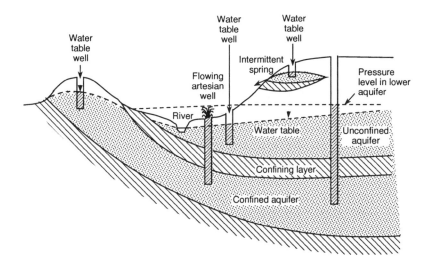

Figure 2.4 Two aquifers are shown in this diagram: an upper one which has a water table; and a lower one which is confined. You can also see how confined and unconfined conditions can exist in different parts of the same aquifer. The water level in the wells into the water table or unconfined aquifer are the same as those in the ground, while those in the confined aquifer are above the top of the aquifer and may flow out at the surface where conditions allow it. A perched water table is supported on a very limited area of low permeability rock. Perched water tables have limited storage and if used for water supplies, are likely to be unreliable

dry weather; whereas springs and wells which obtain their water from the main water table are likely to be reliable, even in the driest of summers.

There is a close relationship between groundwater and surface water since the base flow component of river water comes from local groundwater. Where close contact exists the water table is usually just above river level. Where a stream or river runs at a higher level than the local water table, however, water may be lost from the stream bed and this water will percolate down to the water table and add to the groundwater resources. In some areas river flow will decrease downstream or may even dry up altogether. This is often a seasonal phenomenon controlled by large fluctuations in the water table.

ARTESIAN WELLS

When an aquifer is overlain by layers of impermeable rocks, the pressure of groundwater can be such that the level of water in wells would rise above the top of the aquifer. In such instances the aquifer is said to be "confined". Sometimes this pressure may be sufficiently great for groundwater to flow out of boreholes without pumping; this condition is termed *artesian flow* and both the aquifer and wells which tap it are said to be *artesian*. If you are able to obtain a water supply from a borehole which is artesian you should be able to get away without pumping costs. It is very important to control artesian flows because if you let water run to waste, the artesian pressure may eventually be lost and you will have to resort to pumping to obtain your water supply.

The term "artesian" comes from the Artois province in northern France where the first such wells are said to have been made. Another text-book example of artesian conditions is the London Basin, where the aquifer is chalk outcropping in the North and South Downs. The chalk is overlain by London Clay, which acts as a confining layer, and the difference in ground levels between the Downland recharge areas and the central part of the basin is such to create artesian conditions. During the last century, water from boreholes in central London flowed out at the surface and the fountains in Trafalgar Square were originally fed from this flow. Continuous heavy abstraction over the years has caused the water level to decline and the artesian flows have now ceased.

There are many other examples of extensive artesian basins around the world. In North and South Dakota the main aquifer rock forms the Black Hills and then dips down to the east beneath confining rocks. The water is tapped by artesian boreholes over an area of almost 40 000 km². The biggest artesian basin in the world, however, lies in Australia's Queensland and adjacent parts of New South Wales where a sandstone aquifer provides supplies to artesian wells over an area of around 1 500 000 km². There are other large artesian basins in north Western Australia and in Victoria. Many oases of the Sahara and other deserts owe their existence to groundwater flowing to the surface under artesian pressure.

HOW WATER ENTERS A WELL

Groundwater flow and well hydraulics are topics which follow from the consideration of groundwater in the hydrological cycle, but for the purposes of this book it is not necessary to go into detailed theory. The following description will help you understand the processes which control the flow of water into a well so that you are able to follows the sections on well design and pumping tests in later chapters. Once again, in the context of this section the terms "well" and "borehole" can be used synonymously.

Let us consider what happens when water is pumped from a well. Initially the water in the well will be at the same level as the water table in the aquifer. When the pump is started, water in the well is removed, which lowers the water level in the well. This alters the balance between the water in the well and that in the aquifer. Water then flows into the well from the aquifer to replace that which has been pumped out of the well and restores this balance. This flow causes a lowering of the water table in the aquifer next to the well. As water continues to be pumped from the well, however, the balance cannot be restored. Consequently, more water flows into the well from the aquifer and the effect of reduced water levels gradually spreads away from the well further into the aquifer. The shape of the lowered water table resembles an inverted cone and is often called the *cone of depression* or the *cone of exhaustion* (see Figure 2.5).

At the onset of pumping, changes in well water levels are rapid but slow down and eventually reach a state of equilibrium. The

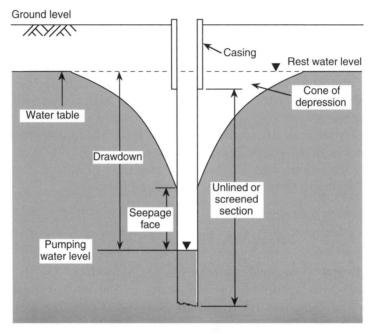

Figure 2.5 Various features of a pumping well or borehole are illustrated in this drawing. This example shows a borehole in a water table aquifer and has a cased upper section. The lower part is likely to be unlined in solid rock aquifers or screened (with a perforated pipe) where support is needed. Pumping has lowered the groundwater levels to form a cone of depression around the borehole. The drawdown is the difference between the water table level measured before pumping started and that measured in the well during pumping. The pumping water level does not represent the groundwater level just outside the well because a seepage or wetted face develops in a borehole. When pumping first starts, the water level falls quickly but the rate of decline gradually slows down until it reaches an equilibrium

difference between the pumping water level and the original rest water level is termed the *drawdown*. The amount of drawdown in a well is made up of two elements: the first is caused by the permeability of the aquifer and the second is caused by the permeability of the well face. The permeability of the aquifer is fixed by nature but the permeability of the well face depends on the design and method used to construct the well.

The cost of pumping from a well depends very largely on the depth to the pumping water level. That factor is controlled by the amount of drawdown and it is important, therefore, that wells and boreholes are carefully designed and constructed to keep drawdown to a minimum.

CHANGES IN THE AQUIFER

As pumping proceeds, the water level in the well continues to fall and the cone of depression gradually increases but ever more slowly. The water flowing into the well is partly made up of water taken out of groundwater storage in the aquifer and a gradually

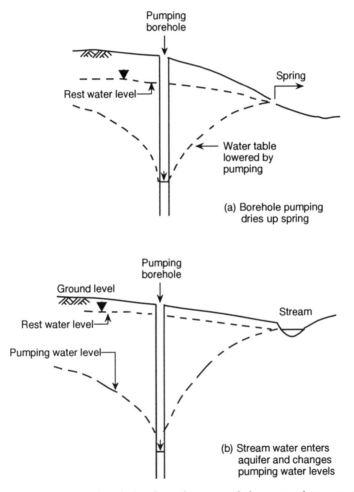

Figure 2.6 (a) A pumping borehole where the cone of depression has extended to a spring. Before pumping started, the water table sloped towards the spring but pumping has reversed this situation causing the spring to dry up. (b) A cone of depression has developed around a pumping borehole which lies near a stream. The changes in the water table have caused water to flow out of the stream into the aquifer, thereby reducing the stream flow. It also means that the shape of the cone of depression has been modified with a smaller amount of drawdown on the side near the stream

increasing amount is derived from the recharge which occurs in the cone of depression. Eventually the whole of the well's yield is met by recharge and the cone of depression becomes stable. If there are any springs or other wells inside the cone of depression they are likely to be affected by the pumping. This effect is caused by the water table being lowered and means that spring flows will be reduced and the yield of affected wells could even be reduced. In the same way that the impact on the water table decreases with increasing distance from the pumped well, the potential effect on springs and other wells is reduced the further they are away from the pumping well.

The radius of the cone of depression depends on the pumping rate and the permeability of the aquifer. Other factors may be significant, perhaps the most important one being the influence of rivers and streams. If a pumping well is situated near a river which is in direct contact with the water table, some of the water pumped from the well could come from the river. This happens when the cone of depression grows large enough to intercept the river. At first this causes water to stop flowing out of the ground into the river and then starts river water flowing into the aquifer. Sometimes this can cause a serious reduction in river flow but it may have the beneficial effect that springs and wells on the other side of the river are not affected by the pumping well or, at least, any impact is reduced. Figure 2.6 illustrates these effects and, as you can see, the cone of depression is rarely likely to be symmetrical, with the shape being determined by many factors such as the geology and the position of local streams and rivers.

3
Choosing the Most Suitable Source

From the previous chapter it is clear that there are a number of different parts of the hydrological cycle which can be tapped for a water supply. It is important to choose the source that is capable of providing the most reliable yield. Whether you decide to use a stream or spring, or pump groundwater from a well or borehole, you will need to make some assessment of how much water you are likely to be able to obtain from the source. In this chapter we consider the different measurements that you will need to assess a water source and the techniques you can use to make sense of them.

RAINFALL

Collecting rainfall to provide a water supply seems to be the simplest system of all. *Rain harvesting* is the term frequently used to describe this sort of water supply and examples are found all over the world. The basic requirements are adequate rain throughout the year and a suitable tank in which to store the water. In areas where the rainfall is markedly seasonal you are likely to have problems in keeping the water drinkable when it has been stored for several months. However, in some locations other potential alternatives may not be viable so if you live in an area with streams that dry up and there are no chances of digging a well you should give serious thought to a rain harvesting system.

RIVER AND STREAMS

In general terms, both rivers and streams usually provide large quantities of water, although it is often necessary to build a reservoir to provide storage so that supplies are still available in times of low river flow. The biggest drawback in using a river or stream is that they are liable to pollution. This risk varies from area to area and is least in remote upland areas. The quality problem is one reason why you are unlikely to get official approval to use water from a stream or river for use in milk production. Similarly, your local public health officers are unlikely to be happy with a water supply from a stream which is to be used for domestic purposes, in the catering trade or for preparing food products.

SPRINGS

The flow of water from springs can be very variable indeed. The most reliable springs are those draining deep-seated aquifers and having little or no surface water component. The water discharging from these springs can be amazing. I recall seeing one which discharges from a major limestone aquifer in central Ireland. The water wells up through the ground in a large number of small flows starting a stream which becomes as wide as an average road in a matter of a few metres. In contrast, most springs which drain perched water tables and those which are supported by flow from aquifers such as some limestones or granite, may dry up after a few weeks without rain. These rocks allow water to drain from them quickly and the resulting springs often have very strong flows shortly after rainfall which may encourage their use for water supplies. But even the best may dry up during the summer and, even if it does not dry up completely, there may be several weeks at a time when water rationing could be necessary.

Some of the springs marked on topographical maps are not really springs at all. Certainly they have a flow of water coming from them and look like springs but, in fact, are actually land drains. This may not necessarily rule them out as a source of water. Indeed, there are many people who have relied on such a drainage system for their water supply for many years, and never had any problem at all, often not even aware that they are not using a spring. One major drawback to this kind of system is that the whole drainage system needs to be properly maintained, otherwise the supply may dwindle or could become polluted.

You really ought to find out if your spring is really a discharge from a land drain so that you can maintain it properly. Many springs have been diverted through pipes to improve field drainage which gives rise to confusion. You can check whether this is the case with your spring by examining the *catchpit* and looking for other signs. A catchpit is the small tank, usually made of stones, bricks or concrete, from which the spring flows. Check to see if the flow into the catchpit is through a tile drain or a stone-lined culvert or drain. These drains may be about 10 cm or 15 cm in diameter. If the water enters through such a pipe it suggests that the supply is from a drainage system but this test is by no means conclusive. You should then look for other signs of the drainage system both upstream and downstream from the catchpit. These drainage systems usually consist of a stone-lined culvert running along the centre of what appear to be dry valleys. The culverts are usually a metre or so below ground and may have catchpits and water troughs incorporated into them from which water supplies are taken. A good walk around the local areas should enable you to find out if your supply is part of a land drain. If it is, you will probably find that several of your neighbours obtain their water from other parts of the same system and you should all co-operate over maintenance and protection of the supply.

It is easy to tell if you have a groundwater spring during frosty weather. Groundwater is always around the average annual ambient temperature which in mid-latitudes is 10–12 °C. In the United States, for example, it has been found that groundwater is generally 1 or 2 °C higher than the local average air temperature. This means that groundwater is over 20 °C in States such as Texas, Louisiana and southern California, around 15 °C from Nevada to Carolina, 10 °C across much of the High Plains and down to 5 °C or less near Lake Superior. This locally stable temperature means that groundwater will not freeze easily, which can be a real giveaway in the frost. As spring water is warm it will enable water plants to thrive round the spring throughout the winter. In contrast, the discharge from land drains is not likely to maintain the same growth. So even if there is no frost on the day you inspect the springs you may still be able to decide on the groundwater element just by looking.

The big advantage of a spring supply (or even a land drain) is that it is made up of groundwater which has reached the surface under gravity flow. With a bit of luck, and perhaps a great deal of good planning, you can utilise such a source and never need to pump. Water for free — well, almost! Pick a spring which is at a higher elevation than your house, cattle troughs and other places where you want water; gravity will bring the water down to where you need it. In this age of increasing energy costs such considerations are important and, if you have a spring, which you may be able to use for your water supply, you would be well advised to investigate its potential before considering other types of source.

A land drain is one way in which groundwater can be brought to the ground surface without pumping. Another more ancient system which has been in use for about 2500 years are the *qanats* of the Middle East. These are gently sloping tunnels which connect a series

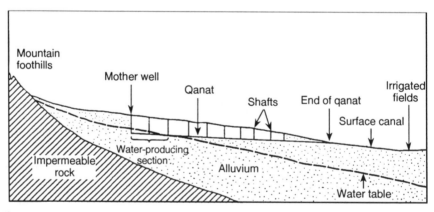

Figure 3.1 A qanat consists of a long, gently sloping tunnel dug through alluvial material which leads groundwater to the surface by gravity flow. They are constructed by a "mother well" being dug to below the water table at the upper end, which is then connected to a series of access shafts until it reaches the surface at the lower end where the water typically flows into irrigation systems. They are found in the arid countries which stretch from Morocco in the west to Afghanistan in the east and can be very large structures indeed. A few have also been built in Spain and parts of Latin America. The largest in the world is at Zarand in Iran and is 29 km long. The mother well is 96 m deep and there are a series of 966 shafts along its length. (Redrawn from Beaumont by permission of the Institute of British Geographers)

of spaced wells or shafts and which break the ground surface at one end. Figure 3.1 illustrates how these tunnels tap groundwater for drinking supplies and irrigating crops. Although believed to be of Iranian origin, qanats are found throughout the Middle East and in other areas of Arab influence such as north Africa and Spain, and via this route into parts of South America. They are called by different names in each country, some of which are listed in the glossary at the end of the book.

WELLS AND BOREHOLES

One of the best indicators of the likely success of a new well or borehole is the presence of a number of successful wells in the locality. This may not be a completely infallible method in areas with variable geological conditions where you may be unlucky enough to own the only field with no water-bearing rocks under it. Where there is a large number of wells or boreholes in an area, a saturation point could have been reached. Pumping from one borehole may affect others nearby or dry up springs. In such circumstances you may have problems obtaining a new supply and have to look elsewhere for your water.

If there are no existing wells or boreholes in your area there is no need for immediate panic as you could be the first person to want to sink a well. Over much of the world, a properly designed and constructed borehole will supply enough water to meet the needs of a household or even a small farm. To some extent all new wells and boreholes are speculative, but the chance of a good supply can be greatly improved by correct siting of the well and having the right design.

WATER DIVINING

Water which flows out of sight underground has been a mystery to most people since time immemorial and it is not surprising that it is surrounded by mystery and legend. In former times springs were often thought to be the dwelling places of local gods and many traditions have developed around these springs — or "wells" as the majority are traditionally called. Close examination of the map for almost any part of Britain, for example, will reveal a holy well or other springs which have been given their own name, linking them with these past beliefs. Many springs are attributed with healing properties or other magical powers and may be part of local tradition. Perhaps some of the best known of these traditions are the well-dressing festivals in Derbyshire and Staffordshire. Even the law has seen groundwater as something mysterious. As recently as 1984 the Ohio State Supreme Court overturned an 1861 ruling that groundwater is "too secret and occult" to be adjudicated by the law. With this close association of magic and water, there is small wonder it extends to people who have a reputation for being able to locate places to dig a successful well. In the UK, such people are called *dowsers* or *water diviners* and in North America are referred to as *water witches*.

It cannot be denied that chance can play a part in well location. Even when the most sophisticated scientific techniques are used, there may be a small element of luck which could make the difference between a "gusher" and a barely adequate supply. In these circumstances it is hardly surprising that water diviners do not only claim to locate water but actually set out and do it! After all, their luck is as good as the scientist's and, when it is coupled with local knowledge and experience of looking for water, given favourable conditions, a dowser may be successful more often than not.

A fundamental scientific principle is that all facts can be demonstrated by repeatable experiments carried out under strictly controlled conditions. So far, water divining — or any other sort of divining for that matter — has not been shown to work when subjected to such objective scientific examination. The American investigator, James Randi has issued several challenges to dowsers in different countries to practise their arts under controlled conditions. It is reported that in each case, the success rate only matched that which could be expected purely by chance. If divining is ever proven, there will be a major revolution in the location of valuable mineral resources ranging from petroleum to uranium and from coal to water, and a lot of unemployed scientists.

There is a lot of circumstantial evidence to support divining, especially in the water field, but I am convinced it can all be attributed to luck and local knowledge. In my experience, those water diviners with the best reputations for finding water operate in areas where a successful well could be dug almost anywhere. When there are failures, there always seems to be a very good reason such as interference from overhead electricity cables or the driller misunderstanding his instructions and constructing the borehole in the wrong spot. It never seems to be the dowser who is wrong! In the ten years since I wrote the first edition of this book the popular interest in dowsing seems to be undiminished, related perhaps to the "new age" movement and revival in all sorts of magic. I have even seen toy dowsing rods for sale in gift shops but would not recommend you use them to locate your water source!

To use practical experience based on a knowledge of the behaviour of other boreholes in the area is usually the best way of siting a borehole. In difficult areas the services of an experienced hydrogeologist may be needed to assess the availability of a water supply on your land. There are many consultants and government departments who will assess the likely success of a proposed new borehole. They base their reports on the geological records and information on other wells, backed up by professional training and experience.

RAINFALL MEASUREMENTS

It is usual for the amount of rain to be talked about as a depth of water, with the average rain for an area being 35 inches or 880 mm, for example. The way in which rainfall is measured in a rain-gauge fits in with this idea. There are a number of different types of rain-gauge used by official weathermen round the world but you need not worry about getting hold of one of the standard rain-gauges, as satisfactory measurements can be taken using a home-made device. Figure 3.2 shows a couple of ways in which this can be done. Remember that you will be recording the amount of rain as a depth of water and this will require you to do some arithmetic. The calculation is quite simple and is explained by the diagram. You should also remember to take the measurements at the same time each day. The British Meteorological Office, for example, has all its rain-gauges measured at 9.00 am GMT. The rain measured at this time fell since you took the last reading so you should record it as the rainfall for the *previous* day. The rainfall you measure on the morning of 23rd May, for example, should be recorded as the

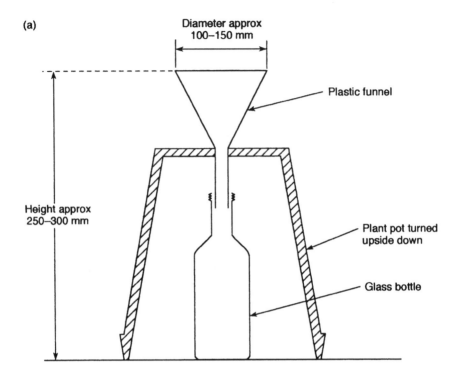

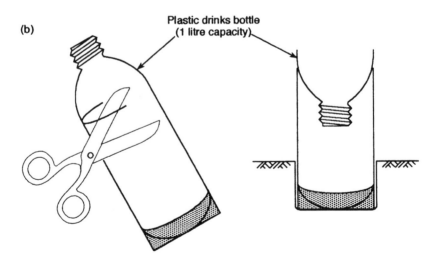

***Figure 3.2**(a) and (b)*

(c)

Rainfall record - May 1994

Date	mm	Notes	Date	mm	Notes
1	-		16	2.7	
2	-		17	1.1	
3	2.0		18	0.5	
4	1.7		19	-	
5	Trace		20	-	
6	0.5		21	4.6	
7	0.3		22	-	Dry and sunny
8	-		23	-	
9	-		24	-	
10	2.0		25	-	
11		Acc. total	26	-	
12	16.5		27	-	
13	7.1		28	10.5	Thunderstorm late pm
14	4.1		29	5.0	Wet in morning
15	Trace		30	6.5	
			31	3.1	
				68.2	

Figure 3.2 You can make rainfall measurements by using an improvised rain-gauge made from readily available materials such as the two examples illustrated here. In (a) rain is caught by a funnel which guides it into a glass bottle. The funnel is prevented from moving by an inverted flower pot which also protects the funnel and bottle from being knocked over. In (b) a large plastic drinks bottle has been cut in two to make both the funnel and collection jar. Take care to make the cut straight so that you get an even top to the funnel. The plastic is slippery so you may need to hold the funnel in place with a piece of sticky tape. This system is very light and will need to be partly buried to prevent it being knocked or blown over. All rain-gauges should be protected so that they are not easily accessible by animals or small children. Rainfall is recorded as a depth of water in millimetres. To take your readings, measure the volume of water collected in the bottle using a measuring cylinder. You are likely to be able to buy these from a hardware store or in a cooking equipment shop. To convert the volume from millilitres (the same as cubic centimetres) to a depth in millimetres, use the following formula:

$$h = v/10\Pi r^2$$

where h is the depth in millimetres, v is the volume in millilitres (or cm^3), and r is the radius (i.e. half the diameter) of the funnel measured in centimetres. Record the value of rainfall you calculate rounded down to the nearest millimetre as shown in (c). It is usual to record the reading you take on one morning as the rain for the previous day. For example, the 2 mm of rain recorded for 3rd May was measured at 9.00 am on 4th May. Make a note of significant precipitation events and if you forget a reading show the total for two days as in the example for 11th/12th May rainfall for 22nd. If you forget to take a reading one morning, split the reading over the days since you took the last one. The diagram also shows how to set out your rainfall readings.

Siting a rain-gauge correctly is important; it must not be in a sheltered spot where it will under-read or in an exposed place where the wind will blow the raindrops away from the top of the gauge. Keep it away from tall trees or buildings by a distance equal to at least twice the height and use your common sense and judgement about exposure. Do not forget to protect it from being knocked over by curious people or animals.

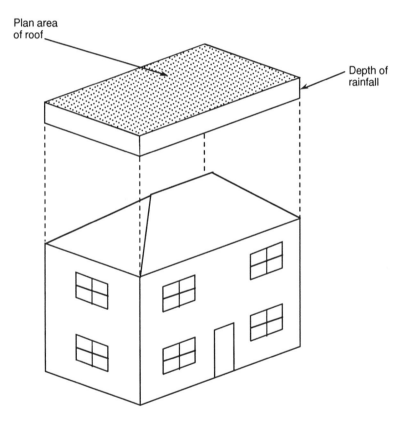

Figure 3.3 The volume of water available for a rain harvesting water supply scheme is calculated by measuring the plan area of the roof (or roofs) which make up the catchment, and then multiplying this area by the depth of rain in a year. Variations throughout the year can be assessed by using the average rainfall for each month. In this way you will be able to work out how much storage will be needed to overcome the dry periods

To assess the yield of a rainfall collection system simply multiply the *depth* of rain which falls during a year by the *plan* area of the roof to give the total available water. Be careful about units, especially if you are using gallons, feet and inches.

You will have to reduce your estimate a little to allow for evaporation losses. These should not be very high, however, as there is virtually no evaporation when rain is falling and the collected rainwater should be protected from evaporation when it is stored in the tank. Relate this volume to your estimated total water needs for a year to find out how much of the total rainfall you will need to harvest. You are unlikely to have a reliable supply if you need to depend on most of the annual rainfall to meet your needs.

It is important to look at the amount of fluctuation from one year to another to see if in relatively dry years there is likely to be enough water. Also carry out a similar exercise on the seasonal variation in rainfall to determine the range of values for the daily or weekly input to your storage tank. Figure 3.3 shows how to manage these calculations.

RIVER AND STREAM FLOW MEASUREMENTS

The flow of most major rivers is recorded continuously by government agencies at specially built gauging stations. Even where flows have not been recorded in this way, it is quite likely that a few spot measurements have been taken at some time over the years. This information is usually available to members of the public, although a small administration charge may be made to cover the cost of things like photocopying. It will always be worthwhile contacting the local office to see if they have any flow records for any stream in which you are interested before you start taking your own field measurements.

Thin Plate Weirs

The most accurate method of measuring the flow of a stream is to use a thin plate weir. All weirs work by restricting the size of the stream channel. This causes water to pile up on the upstream side before passing over the weir as a jet or *nappe*. The rate at which water flows over the weir depends on the height of water above the weir on the upstream side. It is relatively easy to measure water levels and use them to calculate flows. It is even easier with a thin plate weir because, as they are standard shapes, the flow can be looked up from tables.

There are two main types of thin plate weir: a V-notch weir and a rectangular weir. The dimensions and method of operation of these weirs are described in official publications such as the British Standard 3680 Part 4A. The specification given in these documents will ensure accurate measurements to within 1%, but it is not vital for you to be as good as this.

You do not need to buy a weir plate as you can make your own and provided you follow the general guidelines given here for their construction and installation, you may be able to achieve flow measurements with possibly 5% accuracy. You can use sheet metal or even wood to make the weir. It is important to cut the angles as accurately as possible, and to make sure that all edges are sharp and straight and that the upstream face is smooth. If you decide to use wood, choose marine ply which will resist the water and also treat it with a couple of coats of yacht varnish. This will prevent the wood from swelling with the water and breaking up which will make it progressively less accurate. Figure 3.4 shows the general features of weir plates. It should have a lip of between 1 and 2 mm and the downstream face should slope away from the lip at an angle of at least 45° in the case of a rectangular weir. For V-notch weirs this angle must be at least 60°. Flows of up to 60–70 l/s can be measured with a V-notch weir and higher flows can be measured with a rectangular weir of appropriate width.

V-notch Weirs

There are three types of V-notch weir, all with different sized notches. These are a 90° notch which has the width across the top equal to twice the depth, a ½ 90° notch with the width across the top equal to the depth, and a ¼ 90° notch where the width across the top is equal to half the depth. If these dimensions are used it is relatively simple to construct the notch by cutting an appropriate equilateral or isosceles triangle from the plate. You will notice that the angles of the ½ 90° V-notch and the ¼ 90° V-notch weirs are not

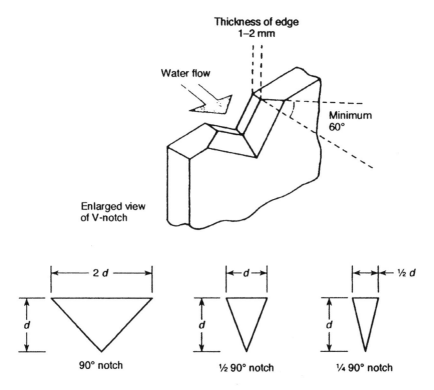

Figure 3.4 Thin plate weirs should have carefully cut edges to ensure accurate readings. This example shows the detail of the edges of V-notch weirs and the dimensions of different-sized notches

45° and 22.5° as you might expect because these weirs get their name from the fact that they record approximately one-half and one-quarter of the flow of the 90° V-notch, respectively. For those who are mathematically minded, the angles are 53° 8' and 28° 4' respectively.

Installing a Thin Plate Weir in a Stream

When a weir plate is installed in a stream channel it is important to select a straight section at least 3 m long. You should make sure that the plate is both upright and at right angles to the direction of flow. Hold it in place with a couple of stakes at each end. To make it easier to install, use a number of sandbags to create a temporary

34

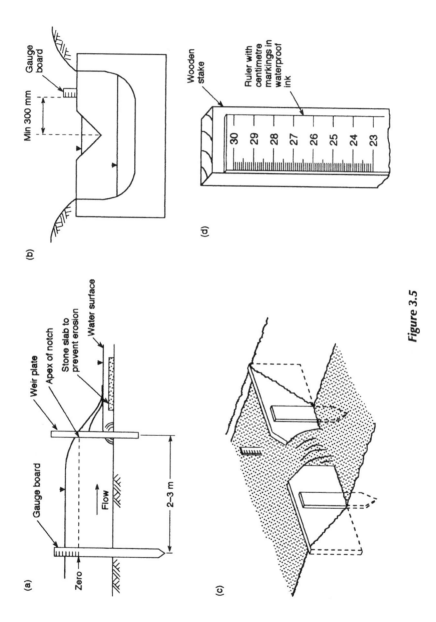

Figure 3.5

dam and enable you to work in relatively dry conditions. It is vital to prevent water from leaking around the edges or underneath the weir. The edges can be sealed with clay, which is best pushed in with the fingers. The level of the crest of a rectangular weir or the apex of a V-notch weir should be set at a height above the stream bed that will ensure an adequate fall on the downstream side. At maximum flow this must be at least 75 mm above the water level in the channel on the downstream side (see Figure 3.5).

Taking the Flow Measurement

As weirs work by controlling the level of the water surface on the upstream side, the flow of water over a weir is related to the depth of water over a weir. If the depth of water is known, the flow can be looked up in the tables, but it is essential to measure the water level in the right place, as shown in Figure 3.5. The relationship between water level and flows over various thin plate weirs is given in Table 3.1.

The biggest drawback to this method of measurement is that the weir will tend to silt up on the upstream side. It is vital to keep it

---

Figure 3.5 Divert the stream or at least control the flow using sandbags, before you start to install a weir plate. Dig out a small trench at right angles to the stream flow to take the plate. Place it in the trench with the sharp edge facing upstream. Make sure that it is vertical and that the top is horizontal by using a spirit level. You may find it best to hold the plate in place using wooden stakes. This operation must be done quickly before the trench collapses. Seal in the plate to prevent leakage which will ruin the measurements, and also eventually wash the installation out. Use clay to make it watertight, pushing it in with your fingers where necessary. Place stones or a concrete slab on the downstream side to prevent bed erosion and make sure that it does not interfere with the flow of water as this will ruin the readings. Install a gauge board on a vertical post 2–3 m upstream of the weir plate. It is best to position it to one side of the channel rather than in the centre. The gauge board should be accurately marked off in centimetre gradations and a cheap wooden ruler can be a ready-made gauge board. If you paint alternate marks in a distinctive colour it will help you read the water level to the nearest centimetre from the bank. The zero of the gauge board needs to be at the same level as the crest of the weir or the apex of the V-notch so check this carefully with your spirit level and a length of straight wood. Take care when you remove the sandbags so that a sudden rush of water does not wash the plate out. Measure the flow by reading the height of water above the apex of the V-notch using the gauge board. Look up this level reading on Table 3.1 to convert it to a flow. Daily measurements should be adequate to assess the quantities of water which are available in the stream catchment

Table 3.1 Flow over thin plate weirs in litres/second. The values given in this table have been calculated from standard formulae. The flow over a 90° v-notch weir is given by the formula $Q = 1.342\, h^{2.48}$ where Q is the flow in m^3/s and h is the head in metres. The flow over a one metre length of a rectangular weir is given by the formula $Q = 1.83\,(1 - 0.2\, h)h^{1.5}$ where Q is the flow in m^3/s and h is the head in metres. For weirs of other lengths simply multiply the flow for a one metre weir by the length of your weir in metres

Head (cm)	V-notch weirs			Rectangular weirs			
	¼ 90°	½ 90°	90°	0.5 m width	1.0 m width	1.5 m width	2.0 m width
1	0.005	0.01	0.02	0.7	1.3	2.0	2.6
2	0.02	0.04	0.1	2.6	5.2	7.8	10.4
3	0.05	0.1	0.2	4.7	9.4	14	18
4	0.1	0.2	0.5	7.5	15	23	30
5	0.2	0.4	0.8	10	20	30	40
6	0.3	0.6	1.3	14	27	40	54
7	0.5	0.9	1.8	16	33	50	66
8	0.7	1.3	2.6	20	41	61	81
9	0.9	1.7	3.4	25	49	74	98
10	1.2	2.2	4.4	29	57	85	114
11	1.5	2.8	5.6	33	66	99	132
12	1.8	3.5	7.0	38	75	112	150
13	2.2	4.3	8.5	42	84	126	168
14	2.7	5.2	10	47	94	141	188
15	3.1	6.1	12	52	104	156	208
16	3.7	7.2	14	57	114	171	228
17	4.3	8.4	16	63	125	188	250
18	5.0	9.6	19	68	136	204	272
19	5.6	11	22	74	148	222	296
20	6.4	12	25	80	159	239	318
21	7.2	14	28	86	171	256	342
22	8.1	16	31	92	183	275	366
23	8.5	17	35	98	196	294	392
24	10	20	39	105	210	315	420
25	11	22	43	110	220	330	440
26	12	24	48	117	235	353	470
27	13	26	52	122	245	368	490
28	15	29	57	130	260	390	520
29	16	31	63	138	275	412	550
30	17	34	68	145	290	435	580

clean so that the water on the upstream side is deep enough for the weir to work. You can do this with a shovel but it will mean getting wet. The back-acting digger on a tractor will be easier and keep you dry, but take care not to disturb the weir plate as you dig. For a rectangular weir the minimum depth of water required is 60 cm. A V-notch which is 30 cm high will need a depth of 45 cm, and 30 cm of water is required for a 15 cm size notch. It is important to remember to clean the section out before the measurement is taken. Do not take the reading until stable conditions have been re-established. This may take several minutes — so be patient!

Choosing the Correct Weir

When selecting the best type of thin plate weir to use you must consider the range of flows that are likely to be measured. Table 3.2 shows the best weir to use for several ranges of flow and a few hints are provided in Figure 3.5. It is important to see how the channel dimensions fit in with the requirements for the various types of weir. Obviously, it is not possible to fit a 2 m rectangular weir into a 1 m wide ditch but, perhaps less obviously, it is just as inappropriate to try to fit a small V-notch into a wide channel.

Table 3.2 Recommended maximum flows for different measurement methods

Method of measurement	Maximum flow for accurate measurement (l/s)
Jug and stop-watch	
1 litre jug	0.1
5 litre bucket	0.5
Weir plates	
¼ 90° V-notch	17
½ 90° V-notch	34
90° V-notch	68
Rectangular weirs	
width 0.6 m	170
width 1.0 m	290
width 1.3 m	380
width 1.6 m	470

Estimating Stream Flow

It is possible to obtain a rough estimate of the flow of a stream using a crude form of the velocity/area method of river flow measurement. This involves estimating the velocity of the flowing water and the cross-sectional area of the stream channel as shown in Figure 3.6. The flow is calculated by multiplying the velocity measured in metres per second by the cross-sectional area measured in square metres. This will give the answer in cubic metres per second and to convert this to litres per second simply multiply by 1000. Other conversion factors are given in Appendix 1.

This result can be used as the basis for a rough estimate of the range of flows in your stream. The maximum depth of water during flood flows will be indicated by pieces of grass or other debris on stream-

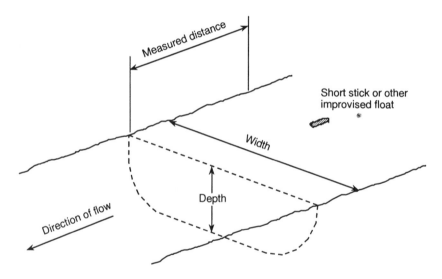

Figure 3.6 Stream flows can be estimated by using the velocity–area method. Choose a length of stream channel which runs straight for about 3–4 m. Use a steel tape to measure the width and depth of the channel at right angles to the flow. If the stream bed is irregular, you should measure the depth at various distances from the bank, and use these measurements to work out an average depth. Multiply these values to get the cross-sectional area of the stream. The water velocity can be estimated by timing a piece of stick, or other suitable small float, over a measured distance of 1 or 2 m. If the water is flowing rapidly it may be prudent to increase this distance to 5 m or even 10 m and you should take the average of several readings. The velocity you need is the average for the stream as a whole. As the water surface moves faster than the majority of the water, it is necessary to make a correction by multiplying your velocity value by 0.7 where the depth is no more than 1 m, or 0.8 if the depth is more than 6 m

side bushes. Compare this height above stream bed level with the depth of water found in the channel when you took your measurements. This will give you some idea of the flow during these flood conditions. Be careful though, as river flow is not directly proportional to the depth of water. For example, if the flood level is twice the depth of the present water level it is not likely that the flood flows will be only twice the flow you have estimated. They may very well be more than four or five times this flow because the cross-sectional area is likely to be proportionately greater, as also is the flow velocity.

Using River Flow Measurements

After the flow of a river has been measured it can be plotted as a hydrograph (see Figure 2.3). This is the best way to see how flows change with time. The lowest flows will define how much water you can rely on being able to take as a direct abstraction. The total flow, over the winter say, will give you an idea of the quantities available for you to store in a reservoir for use during the summer. Obviously you will not be able to use all this water because you would dry up the flow downstream.

SPRING FLOW MEASUREMENT

Flow from a large spring can be measured in the same way as the flow in a stream by using a weir plate. It is important to remember that an adequate length of channel is needed for the weir to work properly and this may mean getting out a spade to dig a suitable channel yourself. Just follow the installation guidelines given above and your measurements will be reasonably accurate. Do not forget that the ½ 90° and ¼ 90° V-notch plates may be more suited to spring flows. Table 3.2 compares various flow measuring methods and will help you decide which technique to use.

Jug and Stop-watch

The simplest way of measuring the flow of small springs is by using the "jug and stop-watch" method. To do this place a small vessel of known capacity below the spring and time how long it takes to fill it. It is essential to have all the flow going into this vessel and to make sure that none of the water splashes out. If you take care, this method is very accurate and potentially the most precise method I describe in this book.

As the name implies, a very popular calibrated vessel for this purpose is the type of measuring jug used in most kitchens. It is best to get one which holds a litre (or two pints) and is made of polythene or metal, although metal ones seem to be difficult to find these days. If you use a jug made of brittle plastic, it will soon crack and spoil the

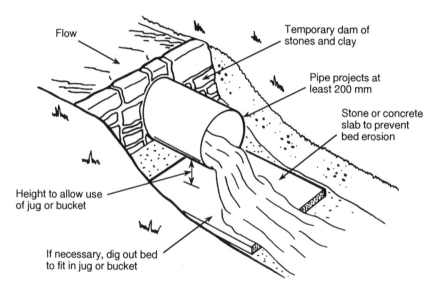

Flow

Temporary dam of stones and clay

Pipe projects at least 200 mm

Stone or concrete slab to prevent bed erosion

Height to allow use of jug or bucket

If necessary, dig out bed to fit in jug or bucket

Figure 3.7 It is usually necessary to modify a spring by diverting all the flow through a short length of pipe in order to measure the discharge. Leave a sufficient gap below the end of the pipe to insert a jug or bucket so that it can stand upright. Build a small dam of clay or stones and concrete as shown in the drawing through which the pipe projects by at least 20 cm. It is important to make sure that there are no leaks so that all the water goes through the pipe. You may have to dig a shallow hole beneath the pipe so that the bucket will fit and do not forget to place a flat stone or concrete slab for the water to fall on to prevent bed erosion. It is important that the bucket is upright when taking a reading, otherwise it will be very difficult to decide when the water level has reached the mark. To get this right ensure that the slab is horizontal and firmly fixed in place. Make sure that you start the stop-watch at the same instant that the first drop of water falls into the bucket and stop it as soon as the mark is reached. Ensure that water does not splash out of the bucket, which can happen especially at the start of the measurement when the bucket is empty. Repeat the measurement at least three times and take the average as the flow. Experience will soon tell you if a greater number of readings are needed

measurements. Try to select a short fat jug rather than a tall thin one as there may not be much space for the jug to fit under the falling water. Accurate measurements need the jug to be upright, so tall ones can be a liability. It is a good idea to make the measuring mark easier to read by painting it or sticking on coloured plastic adhesive. Electricians' insulation tape is ideal for this purpose.

When the spring is running so fast that it fills a jug in less than 5 s, a bigger container such as a 10 litre (2 gallon) bucket should be used. Again, try to choose one which is short and fat and made of robust material. The black polythene buckets sold by most builders' merchants are very good but in all probability you will have to calibrate it yourself. It is quite simple to do this using the measuring jug and clearly marking the 10 litre level with paint or tape.

It will probably be necessary to modify the spring in the way shown in Figure 3.7 before you can go ahead and use your jug and stop-watch. This involves building a small dam so that all the water flows through a pipe and you can use your jug or bucket as appropriate. Remember, if the jug fills quickly, use the bucket — and if the bucket fills rapidly then it would have been better to have used a V-notch weir!

THE YIELD OF WELLS AND BOREHOLES

Before you choose the size of storage tanks and pipes for a water supply system based on a well or borehole, you need to know its yield. You also need some idea of its performance before you can choose the correct pump and the correct depth to install it. It is not only necessary to know how much water can be pumped from the well but you should also find out the change in water levels which occur during pumping. This information can only be obtained by carrying out a *pumping test*. These tests also provide a good opportunity to take samples of water to test its quality. If the borehole has been drilled by a contractor he/she will plan and carry out an appropriate test, but if you are constructing your own well or taking over an existing borehole you may want to do the job yourself.

Choosing a Pump for a Well Test

If the water level is less than 7 m below ground level the tests can be carried out using a surface suction pump. These can be hired easily and the *Yellow Pages* or business telephone directory is probably the best place to find out names of local hire companies. It is always better to have the pump capacity a little too large rather than too small, so if in doubt err on the large side. A suction pump can only lift water to a height equal to atmospheric pressure, which is equal to a column of water about 10 m high, and in practice the maximum lift with the best sort of pump is 8 m, although many can only manage about 6 m (see Figure 7.2). As the water level in the well will go down with pumping, the maximum depth to standing water when you can use one of these pumps will be 6–7 m and not even this amount for less efficient pumps. This sounds a bit limiting but it is possible to use one of these pumps for most dug wells and even some boreholes.

Test Pumping Deep Wells

In boreholes or wells where the water table is deeper than 7 m you can either hire a submersible pump or set up a temporary air-lift pumping arrangement. It is both expensive and difficult to hire submersible pumps so you will probably decide to use air-lift

Figure 3.8 An air-lift pumping system can be made using the sort of equipment shown here. The most important piece of the assembly is the swan-neck. This consists of a length of steel pipe which is bent in a slow 90° angle as shown here. It should be 100–150 mm in diameter. A length of smaller-diameter (25–50 mm) steel pipe is fitted through the curved pipe so that it projects along the central axis. This pipe will be the means of injecting compressed air into the well and should have a strong, water- and air-tight welded joint with the curved pipe. The bottom should project about 300 mm below the base of the swan-neck and have a threaded end capable of taking the remaining air injection pipes. The top end of the pipe should have a compressed air coupling. The swan-neck is fitted to the top of a length of pipes which will form the rising main. These should normally be steel but if the pipe below the swan-neck is shrouded by the well casing, plastic may be used. The air injection pipe projects down the centre of the rising main to about 0.5–1 m above the bottom of the rising main. It is important that a gap of at least 2 m is left above the bottom of the borehole. Normally the gap is much greater than this minimum. The lower 2–3 m of the air injection pipe should be fashioned into a slow taper and be perforated by a large number of small holes. The efficiency of an air-lift pump depends on the depth of submergence below the water level. Under ideal conditions two-thirds of the injector pipe should be submerged during pumping. Use the information in Table 3.3 to calculate the length of the rising main you need; allowing for the drawdown when pumping will reduce the submerged depth

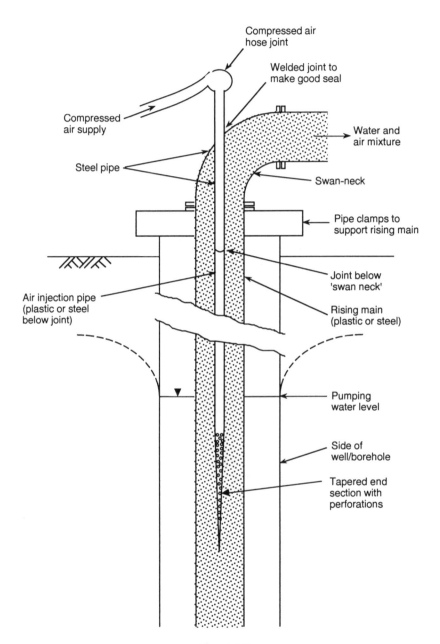

Figure 3.8

pumping techniques. This is not as forbidding as it sounds, provided that you use the right type of equipment and make sure that you follow the safety rules very carefully.

An air-lift pump uses a supply of compressed air from a compressor. The compressed air is forced through the end of a pipe set well below the water level. The air bubbles mix with the water to form a mixture which is lighter than water and so floats upwards. A continuous air supply keeps the air/water mixture moving quickly up the pipe so that water flows out at the surface. It is important to inject the air through a length of pipe which has been perforated with a large number of small holes rather than through a plain unperforated pipe. The large number of holes will ensure that the air and water mix properly and will result in a smooth pumping flow. If the air is injected through the end of the pipe, it will form a series of large bubbles and the water will be delivered in surges. Figure 3.8 shows how to arrange the pipework for an air-lift pump. Table 3.3 lists the depths at which air should be injected for maximum efficiency, depending on water pressures and delivery rates.

Table 3.3   Length of air injection pipes for air-lift pumping are shown in relation to the required lift. For efficient air-lift pumping the injector pipe should be submerged by two-thirds of its length during pumping. The values in this table allow for drawdown but the pumping characteristics of individuals wells will vary and this may mean that a slightly longer pipe is needed

Lift (m)	Length of injection pipe (m)
5	12
10	23
15	35
20	47
25	58
30	70
40	94
50	115

Safety First

It cannot be stressed enough that if you use an air compressor to pump water you must take care that all your connections on the air delivery side are the correct sort of air connectors and hoses, and that they are the right size for the pressures being used. It is very important that all the pipework used to inject the air into the borehole is capable of withstanding these pressures, otherwise they may explode. Although some people have used plastic pipe or copper tubing to inject the air into the water, it is only safe to use them if these pipes are fully enclosed. These materials are unlikely to be capable of withstanding such pressure and nasty accidents can result. It is safest to use steel pipes, although you can use other materials if they are entirely contained within the borehole casing (see Figure 3.8) and cannot endanger anyone should an explosion occur. You should make sure that all the air pipes are tied down to pegs or stakes. Compressors often deliver air in surges under air-lift pumping conditions and the pipe can whip about. If the water is being pumped away from the well through flexible pipes these too should be tied down securely. Safe working practices are also covered in Appendix 2.

Measuring the Pumping Rate

The pumping rate can be measured using a flow meter or one of the techniques described in this chapter for measuring stream or spring flows. If you are pumping using an air-lift you cannot use a meter as it will register both air and water. It is much better to use a weir tank which allows the air to dissipate before the measurement is made. You will probably discharge the pumped water to a suitable ditch or stream. It is important to take the water at least 30 m away from the pumping well to prevent recirculation, and a greater distance if you are pumping at high rates. If the water soaks into the ground and flows back to the well it will give a falsely optimistic yield.

Flows can be measured using the V-notch method but this time the notch is incorporated into a tank as in Figure 3.9. Weir tanks are specialised equipment and may not be easily found. For example, you could isolate part of a dry ditch with sandbags at one end and a V-

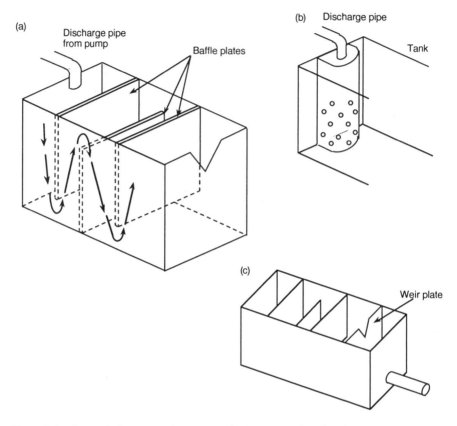

Figure 3.9   Pump discharges can be measured using a weir plate fitted to a tank at least 2 m long. The notch may be cut into one end of the tank so that water can flow over it out of the tank (a). Alternatively, the weir plate is fixed inside the tank with the water being discharged through pipes which gives more control over where the water goes. The water is pumped into the end opposite the notch and then flows over some baffles which will make sure that the flow is laminar by the time the water gets to the V-notch. You can use a wooden board or perforated metal plates to make the baffles. Even a 25 litre drum with holes knocked in it will work if the discharge pipe goes into it (b). It does not matter what is used as long as the correct flow conditions are achieved. The system operates in exactly the same way as a weir plate in a stream. In this case the water level is measured closer to the weir plate but accuracy is not lost as the tank is relatively wide compared to the stream installation

notch at the other to turn it into a temporary weir tank. Alternatively a purpose-dug hole in the ground can be used provided you can get rid of the water. In permeable ground, line the hole to prevent water soaking away before it is measured. Follow the general principles shown in Figure 3.9, whichever method your genius for improvisation suggests!

The "jug and stop-watch" method can be used provided that the pumping rate is small enough. Just follow the procedures described above and do not forget to take repeat measurements. If the pumping rate is too great to use a bucket, a 25 litre (5 gallon) drum or even a 200 litre (45 gallon) oil drum could be used (but make sure that it is clean to avoid pollution). Choose a bucket or drum that will take at least two or three minutes to fill. Even with these large vessels it is still worthwhile to check the calibration and mark off as with a jug or bucket. The main problem you are going to have is emptying them once they are full. If you wish to remain dry, make a hole near the bottom and fit it with a removable bung, but watch your feet when you pull it out! Once again, take the average of several readings and repeat your measurements every half hour or so to make sure that the pumping rate remains constant.

Flow from an Open Pipe

There are a couple of methods for assessing the flow of water discharging from an open pipe which are sometimes used in pumping tests but could be used in other situations. Measurements are taken of the shape of the jet of water coming out of the pipe. By using these results and the diameter of the pipe, the flow of water can be obtained from standard tables. Figures 3.10 and 3.11 show how to take the measurements and Tables 3.4 and 3.5 give the flow rates.

Water Level Measurements

There is usually a lot of room in a dug well to allow water level measurements to be taken. However, there is likely to be very little room to take these measurements in a borehole and the probe can easily get tangled up round the pump. To make matters worse, in a borehole the restricted room will mean that the water surface will be surging very violently during pumping. Figure 3.12 shows how to protect the measuring instrument in a borehole, by installing a tube in the borehole to lower your probe through.

There are various commercial instruments available for water level measurement in wells and boreholes but these are fairly

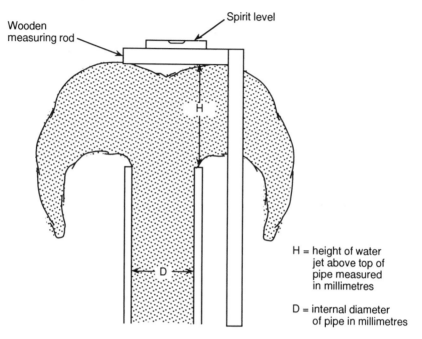

Figure 3.10 You can assess the rate of discharge from a vertical pipe by carefully measuring the height of the water column above the top of the pipe and then using the values given in Table 3.4 for the appropriate pipe diameter. It is usual to have a square made of two pieces of wood at right-angles to help you take the measurement

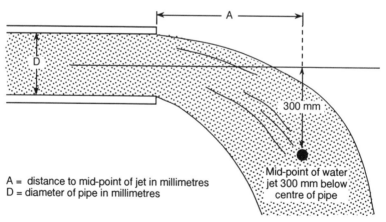

Figure 3.11 To assess the discharge from a horizontal pipe, measure the horizontal distance from the end of the pipe to a point which is central in the water jet and lies 300 mm below the centre of the pipe. It is usually easier to measure from the top of the pipe, subtracting half the pipe diameter from the 300 mm distance. Use Table 3.5 to look up the flow making sure that you use values for the correct pipe diameter

Table 3.4 Discharge from a vertical pipe in litres per second (see Figure 3.10)

Height of jet, H (mm)	Nominal diameter of pipe, D (mm)					
	50	75	100	125	150	200
40	1.4	2.7	4.3	5.3	6.9	10
50	1.6	3.4	5.8	7.5	10	14
75	2.1	4.6	8.1	12	16	24
100	2.4	5.5	9.7	14	20	33
125	2.8	6.2	11	17	24	39
150	3.0	6.9	12	19	27	46
200	3.5	7.8	14	23	32	56
250	3.9	8.8	16	25	36	66
300	4.3	10	18	28	40	72
400	5.0	11	20	32	45	84
500	6.3	14	24	36	56	105
750	7.8	17	30	47	68	125
1000	9.0	20	34	54	80	145

Table 3.5 Discharge from a horizontal pipe in litres per second (see Figure 3.11)

Distance A (mm)	Nominal diameter of pipe, D (mm)					
	50	75	100	125	150	200
150	1.3	2.9	5.0	7.8	11	20
175	1.5	3.4	5.8	9.1	13	23
200	1.8	3.8	6.6	10	15	26
225	2.0	4.3	7.4	12	17	29
250	2.2	4.8	8.3	13	19	33
275	2.4	5.2	9.1	14	21	36
300	2.6	5.8	10	16	23	39
400	3.5	7.5	13	22	31	50
500	4.4	9.6	16	26	39	65

NB Pipe *must* be full.

expensive. The type of instrument in everyday use by professionals consists of a twin-core cable incorporated into a measuring tape. At one end there are a couple of electrodes and when both touch the water they complete a circuit, making a light flash or a buzzer sound. There are a number of makes on the market; an example is shown in Figure 3.13.

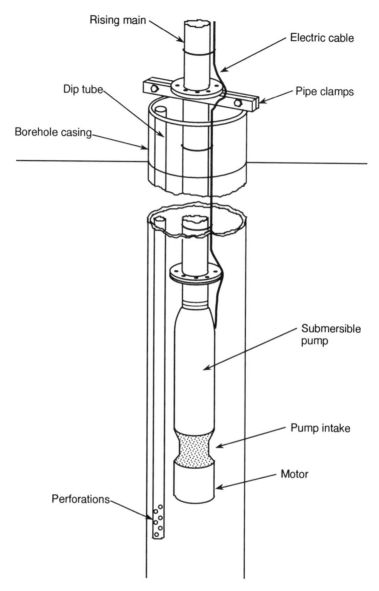

Rising main

Electric cable

Dip tube

Pipe clamps

Borehole casing

Submersible pump

Pump intake

Motor

Perforations

Figure 3.12 A typical borehole installation. A dip tube has been installed to help take accurate measurements of the water level. The bottom of the tube should be positioned below the pump intake as shown and it is important to make sure that the bottom 3 m or so of the tube are perforated. Small holes can be made using an electric drill and they should be no more than 100 mm apart. Plastic pipes are probably the easiest to use because they are light. Solvent-jointed uPVC pipes can be made up at the surface and then hung in one length from the top once the glue has properly set. Take care to keep the pipe in a gentle or slow curve as you feed it into the top of the borehole. The diameter of the tube should be enough to allow the water level instrument to go down it with about 10 mm space all around

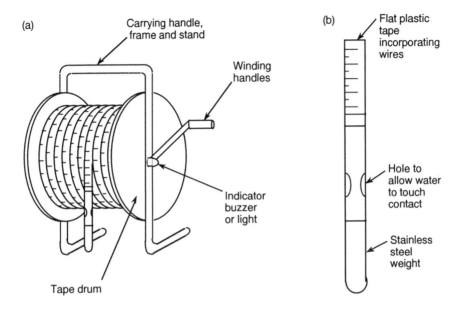

Figure 3.13 There are several companies who manufacture water level probes for using in wells and boreholes. The type illustrated here (a) consists of a plastic tape some 30–100 m long, graduated in metres and centimetres and wound on a drum. A hollow, stainless steel tube with a weighted end is attached to the end of the tape and two thin wires are incorporated into either side of the tape (b). One wire is connected to the steel tube and the other to a needle which is fixed inside it packed in insulating material. When the needle point touches the water surface a circuit is completed which activates a buzzer or light. The measurement is made by "feeling" the water surface (i.e. moving the tape up and down by a short distance) holding the tape firmly between finger and thumb. The measurement is recorded at the point where the light just goes off or the buzzer ceases. Some manufacturers use round section cable instead of tape. These are cheaper to buy but are much less convenient to use, especially during a pumping test

You could make up one of your own using the sort of electrical cable sold for doorbells, provided it is weighted to make it hang straight. Make sure that the two electrodes cannot touch accidentally and that the only way a circuit can be completed is when both are under water. You will have to use a simple electronic circuit if the cable is more than a few metres long. Figure 3.14 shows you how. Before you use the probe in the borehole, test it in a bucket of water. Divide the cable in metre measurements, using insulation tape to form the markers. When using the instrument make sure that the circuit is not completed by the electrodes touching a metal pipe.

Whichever type of instrument you choose, it is used in the same way. Measurements should be taken by lowering the probe into the

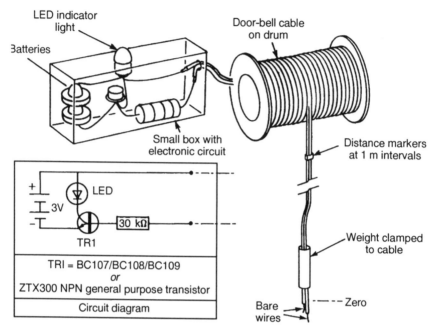

Figure 3.14 A home-made water level probe can be made using the sort of twin-core cable sold for fixing door-bells. Bare the wires at one end but make sure that they are unable to touch by having them at different lengths as shown. Firmly fix a weight to the end of the wire to pull it straight when in use. Fix markers at 1 m intervals along the tape. The other ends of the wires are connected to a small electronic circuit as shown in the diagram. Fix the circuit in a small plastic box to keep the electronic components clean. The probe is used in the same way as a commercial dipper with a steel tape used to measure from the nearest distance marker. Use the one *below* your fingers and add your measurement to the value of the marker

water until the signal comes on, and then pulling it slowly out of the water until the signal stops. This is repeated a few times so that you can "feel" the water level. Use your fingers to mark the tape against a fixed datum point when the light goes off or the buzzer sounds. If the marks on the cable are at metre intervals the distance from the closest one can be measured using a steel tape to give a reading to the nearest centimetre.

An old-fashioned method of measuring water levels, which was commonly used at one time, is the "wetted tape" method. This involves weighting a piece of string and rubbing it with coloured chalk. The string is then lowered into the well or borehole until part of it has been immersed in the water. It is then removed and laid in a straight line on the ground. The depth to water level is measured with a tape using the point where water washed off the marker chalk to indicate the level of water. This method is rather cumbersome and slow to use but can be improved a little if a weighted plastic measuring tape is used instead of the string. The wetted tape method will work at depths of up to 25 m or so, but requires the

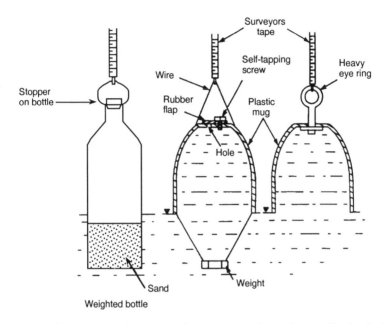

Figure 3.15 Three types of improvised probes are shown here, all of which are designed to hang on a measuring tape or a piece of strong string marked off in feet or metres. The best sort of tape is that used by surveyors, is made of plastic and is usually 30 m long. The tape is rolled into a circular case which prevents it becoming tangled. One of the "instruments" is a small empty screw-top bottle which once held a soft drink. It is weighted with sand so that it will float upright and the top is screwed on before the bottle is firmly attached to the measuring tape. The instrument needs to be calibrated before being used by lowering it into a bucket of water until the weight on the tape lessens and the bottle begins to float. The distance between the zero on the tape and the water level should then be measured and this amount then added to all measurements to give the correct depth to water.

Two variations of "ploppers" are shown on the right. Both are made from a plastic handleless cup or mug (or even an old can) which is weighted so that it will sink when it hits the water. The mug is attached to the tape in such a way that when it is pulled out of the water the flooded cup is inverted and comes out of the water bottom-end first. Atmospheric pressure will keep the water in the cup until the rim leaves the water surface. At this instant the water flows out of the cup and the weight on the line is suddenly reduced enabling the water surface to be "felt". Once again the device will have to be calibrated in a bucket of water to ascertain the constant error in the tape reading

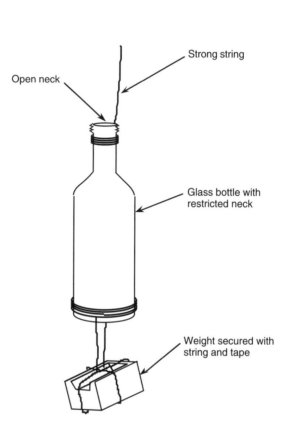

Strong string

Open neck

Glass bottle with restricted neck

Weight secured with string and tape

Figure 3.16 Water samples can be taken using an empty beer or lemonade bottle. It needs to have a narrow neck and be made of glass so that it can withstand high water pressure. It should be thoroughly cleaned and securely tied to a weight of about half a kilogram to make the bottle sink. The neck of the bottle should then be tied firmly to a length of strong string wound onto a reel or piece of stick. Calculate the depth where you want to take the sample and run out a length of string to match it. To take the water sample, lower the bottle into the well as quickly as possible until your pre-selected depth is reached. As the bottle sinks, the increasing pressure will cause the air trapped inside it to bubble out through the narrow neck and prevent a significant inrush of water until the bottle reaches the pre-set level. You should wait long enough for all the air to bubble out of the bottle before pulling it up. It is important to lower the bottle quickly but at the same time to take sufficient care not to drop it to the bottom of the well or break it by hitting the sides. If you want to take a sample from a depth greater than 5–6 m below the water surface, reduce the size of the hole. Leave the bottle cap in place and pierce it with a nail to make a small hole about 2 mm across. This will slow down the rate at which the air leaves the bottle and will make sure that most of the water comes from the level you want

operator to know the approximate depth to water to make sure the tape is submerged.

There are a number of different home-made probes which all work on the principle of weight changing on a line. Figure 3.15 shows several types which are all made from easily available materials. You must be careful when using this type of measuring device as they generally have a constant error which must be added to all the readings you take.

Sampling Water from Wells

The easiest time to obtain samples for chemical analysis is when the well or borehole is being pumped. This also has the advantage that the sample will be representative of the water you will be pumping into the supply. If there is no pump in the well, then you can get a sample using a weighted bottle (see Figure 3.16), unless of course you can borrow a more sophisticated device from one of the professionals in the field. It is important to make sure that samples are taken from below any impermeable casing, that is from the screened or open section of the well. Water remains static in the top cased section and it is usually different from water lower down. A more detailed discussion about water quality and treatment is given in Chapter 6.

PLANNING A PUMPING TEST

Usually there are three things you may want to find out from a pumping test. These are the safe yield of the well, the pumping characteristics of the well, and whether or not pumping from the well will cause problems with other people's supplies. You are likely to need to know the first two facts to design a reliable water supply system, and the third will be required should your new well or borehole need to be licensed. Fortunately, most of this information can be obtained at the same time and from the same tests.

The same sort of measurements of water levels and pumping rates are taken in all three types of test. When the pump is first switched on, the water level in the well or borehole will change rapidly. Measurements of the water level need to be taken very frequently at first; they may then be tailed off gradually. It is usual to take one measurement every 30 s for the first 10 min, then each minute for the next 20 min, and every other minute for the next 30 min. After an hour of frenzied activity you can relax a little as measurements are only needed at 5 min intervals for the next half hour and then at 10 min intervals for the following half hour. By the end of the third hour measurements are needed only two or three times each hour, and within five hours or so, only one reading is required each hour. Once the frequency of readings is reduced it is tempting to hurry off for a coffee or use the loo. Beware of missing the next reading; if you do, note the actual time it was taken. If the test is to go on for several days, water levels are measured only two or three times a day. If information is needed about the rate that water levels recover once the pump has been turned off, measurements are taken at the same sort of frequency. That is, readings are taken twice a minute to start with and are then tailed off as before. Table 3.6 gives a summary of the frequency for water level measurements.

Pumping tests are used to answer questions about a borehole's long-term yield, such as how much water you can get out of it, will you need to pump continuously and will there be enough water for your needs? The other questions answered by these tests relate to the pumping efficiency of the well or borehole.

Over the years, the screen or rock face through which water flows into a well may slowly clog up. This problem has a number of possible causes. Sometimes, the flow of water towards the well draws finer material from the aquifer which may bridge across pore spaces rather than pass into the well. Perhaps the most common way for borehole yield to deteriorate is for there to be a build-up of chemical deposit on the well face rather like the furring in a kettle in hard water areas. As water flows into a well or borehole from different levels in the aquifer it may undergo a change in pressure. This can upset the equilibrium of dissolved minerals in the water, causing them to be precipitated on the well face. Wells and boreholes can also be clogged by bacterial growth. You may think it strange, but there are a large number of types of bacteria which, if they are present in the ground, may thrive in the environment of your well. Some of them can produce a thick, glutinous slime as they grow, which can easily block up the well face. Recent research has shown that bacteria which occur naturally in the ground may also play an important part in the chemical precipitation process.

Table 3.6 Frequency of water level measurements from pumping tests

Time since start of pumping	Minimum frequency of reading
0–5 min	every 30 s
5–10 min	every 1 min
10–20 min	every 2 min
20–60 min	every 5 min
60–100 min	every 10 min
100 min–4 hours	every 15 min
4–8 hours	every 30 min
8–18 hours	every 1 hour
18–48 hours	every 2 hours
48–96 hours	every 4 hours
96–168 hours	every 8 hours
168 hours to the end of the test	every 12 hours

The information in the table is typical of the frequency of measurements used in the UK over the past three decades. The rate of water level change is rapid at the start of the test and then gradually slows down. This means that very frequent readings are needed early on, with the minimum time between readings increasing as the test proceeds. Very few if any pumping tests are carried out with all the readings taken spot on each time. If you are late for any reading, note the actual time you took the measurement. Do not forget that you will have to start at the beginning each time you change the pumping rate. As a result you will be taking very frequent readings all the way through a step test, as each step may be for 100 minutes for example. Some people prefer to take measurements at least every hour, even during the later stages when a lesser frequency would be acceptable. If it is convenient to take extra readings, do so. There is no harm in having lots of data as long as it is of good quality.

The deterioration of a borehole's performance can be judged by ascertaining its *specific capacity*. The specific capacity of a well or borehole is the quantity of water that can be pumped out of it for a unit drawdown in water level. It is measured in peculiar units such as gallons per hour per foot or litres per second per metre. If you establish the specific capacity of a well soon after it has been constructed, these early measurements can be compared with future pumping performance to check on pumping efficiency. When the efficiency of a well goes down, pumping costs go up! Sometimes they may double or even treble! Details of how to cure these problems are given in Chapter 8.

Calculating Specific Capacity

Specific capacity is determined by carrying out a stepped pumping test during which the well or borehole is pumped at three or four increasing rates for fixed periods of, say, two hours. During each step the pumping rate is kept as constant as possible and water level changes are monitored.

So that you can choose the various rates you are going to use in a step test, try out your pump and get a feel for controlling the pumping rate before you carry out the test proper. When using either a suction pump or a submersible pump, control the pumping rate by using a valve positioned in the pipe near the well head. If you are using an air-lift pump, variations in pumping rate can be achieved with a valve or by controlling the compressor. Whichever method is adopted, each pumping rate should be related to a certain number of turns on the valve or to the rate that the compressor is running. This preliminary test should be carried out the day before the step test proper so that water levels have time to fully recover to static conditions.

The way to measure the specific capacity of the well from a step pumping test is shown in Figure 3.17. The specific capacity curve can be used to estimate the drawdown for any particular rate of pumping and is particularly useful in selecting the correct submersible pump. It can also be used as the basis of comparing borehole performance over the years to see if clogging has started. It is surprising how often declining performance is blamed on pump wear, falling water levels or a new abstraction in the area rather than being attributed to the well face clogging up.

Yield Tests

You can get an idea of the reliable yield of your well by pumping at a constant rate. This is a straightforward "suck it and see" exercise, in which the borehole is pumped continuously for anything from a few hours to a day or two. During this period water levels and pumping rates are monitored once again. Hard pumping for 48 hours is usually enough to establish the reliable yield of a new well or borehole but pump for longer if the pumping water level is still falling. The quantity of water pumped from the well during this

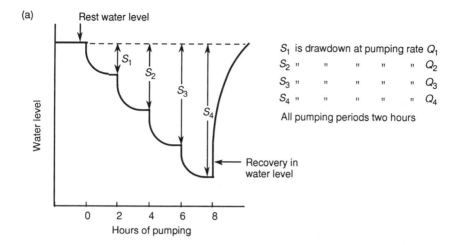

(a) Rest water level

S_1 is drawdown at pumping rate Q_1
S_2 " " " " " Q_2
S_3 " " " " " Q_3
S_4 " " " " " Q_4

All pumping periods two hours

Recovery in water level

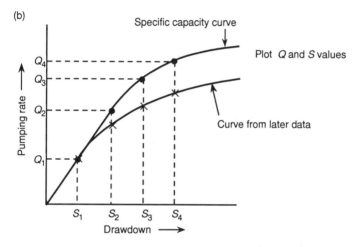

(b) Specific capacity curve

Plot Q and S values

Curve from later data

Figure 3.17 The specific capacity of your well is measured by pumping at several increasing rates for equal periods (a). Measure the water level in the pumping well at the frequency shown in Table 3.6, and do not forget to start again with half-minute readings after each change in pumping rate and also when you turn off. In the example shown here the test has been pumping at four rates for two-hour periods followed by a recovery period. To analyse the data, plot a graph of pumping rate (i.e. yield) on the y-axis against drawdown on the x-axis as shown in (b). This produces a specific capacity curve or yield/drawdown curve for the well, which is used to choose the correct submersible pump. It is also used periodically to compare borehole performance. To obtain the specific capacity of the well from a step pumping test, plot a graph of pumping rate against drawdown. The graph shows two specific capacity curves, the first from the test carried out soon after the borehole was drilled and the second from a similar test carried out a few years later. The difference between the two curves indicates that less water is now obtained from the well for any given drawdown, particularly at the higher rates of pumping suggesting that the well-face has become clogged. This graph can be used to estimate the drawdown for any particular rate of pumping and can be compared with manufacturer's pump performance curves to select the most appropriate submersible pump

period will usually exceed the normal daily water requirement several times over. This will give you the confidence that your borehole will be able to meet all your water needs. The test will also provide operational experience to enable you to plan the plumbing end of your water supply system.

Pumping tests on shallow dug wells often only need to be carried out for a relatively short period. Indeed the test may be limited by the fact that the well can be pumped dry in only an hour or two. If this is the case with your well, adjust the pump so that you can keep it pumping continuously. This may mean feeding part of the pumped water back into the well if it is difficult to control the pump. If you try this system make sure that you put a tee-piece in the pipe *before* the meter or weir tank so that you only measure the actual net quantity pumped out of the well. During the yield test, make sure that the well is pumped much harder than you will ever need to pump during its ordinary operation.

Aquifer Tests

When pumping from a well or borehole requires an abstraction licence, the abstraction control authority will usually require a pumping test to be carried out before any licence is granted. This is to provide them with enough information to work out the impact of your new abstraction on the local groundwater resources and existing abstracters. The authority will issue a consent for you to carry out these tests (see Chapter 9) and will specify exactly how they want you to go about it.

To obtain information to support your licence application, you will be required to locate all the wells, boreholes and springs which are in use within a specified distance from your well. This will vary according to the quantities you hope to abstract and will range from 500 m to perhaps a couple of kilometres. To find these sources you will have to go round and see everyone who lives within the specified area and ask them where they obtain their water supply. You will be expected to monitor these sources during the test period. This will begin a week or two before pumping starts to obtain background information and will continue for a similar period after pumping has stopped to provide enough time to see any effects of your abstraction on neighbouring sources.

You will need to measure the flow from springs and the levels in wells and boreholes using the techniques described in this chapter. The licensing officials will make sure that you are taking the correct measurements and carry out enough checks to pick up any errors you may be making. This all sounds like a great deal of effort to go through to obtain an abstraction licence, but although it may take rather a long time to complete, it is not really complicated. Usually, the licensing authority staff will carry out a brief desk study to see what sort of problems there may be in granting you a licence. This exercise is usually done before a consent is granted to allow you to drill the borehole, so that the terms of the consent can be worked out. If serious problems are anticipated which could result in your abstraction licence being turned down, they will probably refuse you permission to construct your well in the first place. The chances of you going through all this performance and having no licence at the end of it are really quite small.

4

How Much Do I Need?

There is no point in looking for a new source of water or starting to design a new supply system until you have worked out how much water you actually need. The various uses you have for the water will determine the amount you require on an hourly, daily and yearly basis but these quantities will vary greatly from one user to another. You may need a constant flow if you are in the fish farming business, but if you want to supply your home then a mere trickle will probably do, provided it is enough to keep your storage tank topped up. This chapter covers the ways in which you can calculate the quantity of water you will need.

It is important, however, to resist the temptation to overestimate your requirement. It is all too easy to calculate your needs and then double it "for luck", but if you do this you may find it difficult to locate a source of water which will yield these large quantities. If you are successful in constructing a source which will provide this large amount of water, it will certainly have cost you more. Also, it is likely to be much more difficult to obtain an abstraction licence if one is needed and there may be other problems caused by overestimating your requirements. I am not suggesting that you should try to cut things down to a minimum as it is prudent to have some safety margin, but do not be too cautious and cause yourself problems. Table 4.1 gives typical water requirements for a wide range of uses to help you estimate your water needs.

Table 4.1 The water requirements for a number of different uses are shown in this table. Typical daily requirements are listed and maximum flow rates are given to help you decide the capacity of the supply pipes

Use	Daily consumption (l)	Maximum flow (l/min)
Domestic (per person)		
drinking and cooking	4	1
washing and bathing	45	7.5
flushing toilet	50	10
cleaning and washing up	14	2
outside uses	8	2
garden watering	900	20
car washing	100	20
Dairy cow	150	10
Cattle	45	5
Calf	25	2
Horse	50	5
Pig (hog)	20	2
Sheep	10	2
Goats	10	2
Laying hens (100 birds)	30	3
Non-laying hens (100 birds)	20	3

DOMESTIC SUPPLIES

Water is needed in the home for a host of things besides drinking. In addition to providing the essential ingredient in tea and coffee, water is needed for washing the cups and for the inevitable flushing the loo and washing your hands. Water is also needed for cooking, baths, washing clothes, watering the garden and washing the car.

The average adult uses some 2 litres each day (that is about 4 pints) for drinking and cooking in temperate countries like Britain. We all know that we drink more in hot weather and in hot countries the consumption may increase to 3 or 4 litres. Around 60% of what we drink is consumed as tea, coffee or cold drinks like orange squash, and the remaining 40% is made up of bottled drinks including milk, or a quick one at the local bar. It is reckoned that each one of us uses about 20 litres per day for drinking, cooking and personal hygiene. This works out at over 7000 litres in a year! It can be expected that each person flushes the toilet at least two or three times a day. The volume of water needed for that is about 13 000 litres in a year.

Every time you have a bath you use about 90 litres. If this sounds too high, go and have a look at the size of your bath and you will see it is not necessary to have much more than 15 cm of water in the bath to use this quantity of water. Showers use less water than baths, probably only one-quarter of the quantity being needed, but people with showers tend to use them more often and so the overall water consumption is about the same.

Washing clothes is demanding on water consumption. It can be reckoned that each load uses about 130 litres although an automatic laundry machine may use more. The annual water requirement for washing the clothes of a family could be well over 20 000 litres and perhaps almost twice this quantity if there are several young children. When you wash the dishes it takes about 10 litres each time, which is about 30 litres every day. If you use an automatic dish washing machine, the water consumption will go up, perhaps to as much as 60 litres per cycle. Water is also needed for things like car washing, swilling down the back yard and watering both the lawns and the vegetable patch. Some gardens can have a high water requirement if, for example, you have a pond. Water gardens are becoming more popular and need a supply to compensate for evaporation losses during hot dry spells. If you have this problem do not fall for the temptation of using mains water even if the local water company allows it. Mains water is likely to contain chlorine as a disinfecting agent and this will kill off your fish or the frogs, toads and newts which live there. The dissolved minerals in mains water which are not harmful to drink may provide a good food source for algal growth. A combination of the hot weather and mains water is likely to swamp your pond with algae.

Surveys of water consumption show that the volume used varies considerably even in apparently similar households. People on their own use about 120 to 130 litres per day, whereas each person in a family of six will only use about 70 to 80 litres. Logically this must be right, as combining some activities reduces consumption. It takes the same amount of water to wash dishes for one person as it does for two or three.

Domestic water consumption varies throughout the day and usually peaks during the morning when people get up, have a bath or shower, make breakfast coffee and then wash up the pots. Clothes may be laundered during the morning adding to the peak demand. Unless you eat your main meal at midday, there then tends to be a lull until the early evening rush when cooking starts, followed by families getting ready for bed.

It is important to consider what your peak rate use of water is likely to be so that you can either build a large enough storage tank into your supply system or choose a pump which will deliver water fast enough. If you are planning to store your water in a tank and allow it to flow to your taps under gravity you need only think about the size of the tank and the pipes feeding from it. The need to match the size of the pump with peak flow rates only applies where you are pumping to a pressure tank. Chapter 7 covers these aspects in more detail.

The maximum rate will arise when all — or at least a realistic number of — water appliances are in use at the same time. Clearly the quantities involved will vary from one household to another, so we will briefly consider the maximum water use of a large family. Assume that someone is taking a bath or shower while a meal is being prepared in the kitchen and at the same time the dishwasher is in use. If two people also visit the bathroom (assuming that there are two!) and wash their hands under running water, the peak flow rate will be around 6 gallons/minute or half a litre/second. Remember that this peak rate is only needed for a short time and will not increase the overall daily use.

To work out how much water your family is likely to need each day, simply use 130 litres per person. This will allow 520 litres per day for a family of four which is enough to cover for peak demands. If you entertain a lot perhaps you should add another 100 litres or so then you are not likely to go short. Do not forget that if you have a swimming pool you must estimate the quantities you need for this separately from your domestic requirements. You are only likely to want to refill it occasionally but you will have to top up losses caused by evaporation on a regular basis. In calculating the overall figure bear in mind that you should allow enough water for fire fighting. This does not require good quality water and many people build a duck pond to provide an emergency reservoir.

AGRICULTURAL SUPPLIES

An adequate supply of good quality, clean water is fundamental to modern stock and dairy farming. The first thing to decide before designing a water supply system for the farm is how much water will be needed. This figure can be calculated on the basis of the number and type of stock on the farm. There are accepted figures for the water needs for various types of farm animals. These quantities have been used by the agricultural community and the water industry in a number of countries for at least 30 years and have been found to give a good indication of the size of supply required.

Large animals such as cattle and horses drink up to 50 litres per day. Of course they only get up to the full amount on hot summer days. Dairy cows may drink more than 50 litres per day, the total amount depending to some extent on their milk yield. In winter, when they are being fed large amounts of dry feed, cows drink more than at other times of the year. The quantity, which is in proportion to both the amount of dry feed and the milk yield, can exceed 100 litres per day in some circumstances.

Small livestock have a relatively small water requirement. For pigs or hogs, allow 15 litres per day, whereas sheep need up to 7.5 litres per day. Poultry should have a supply of 25 litres per day for every hundred birds.

If you keep dairy cows you will need significant quantities of water for washing and cleaning the dairy. Milking machines need cleaning after use, udders must be washed, and the floor in the dairy and collection area will need to be kept clean. Water is also required to keep the milk cool while waiting for the milk collection lorry. The total water requirement per dairy cow is around 150 litres per day, which includes sufficient water for all these uses.

You can use these rule-of-thumb figures to calculate the needs for your farm on the basis of the stock you keep. These quantities are on the generous side and allow for wastage so there is no real need to add a large safety margin when you are working out how much you need.

IRRIGATION

Crop irrigation is a very important factor in the world's food supply as it enables crops to be grown in dry areas or during periods when otherwise there would be crop failures because of the lack of rain. There are two basic types of irrigation: flood irrigation, where water is poured onto the roots of crops either from channels or pipes; and spray irrigation where water is pumped through pipes into a sprinkler system to produce artificial rain. Spray irrigation is generally more popular but has the biggest demand for water as a significant amount evaporates before if can soak into the ground and reach the plants' roots.

Extremely large quantities of water are needed for the spray irrigation of crops. To give you an idea of how much may be involved, a holding of 16 hectares is likely to require 34 000 litres per hour and 410 000 litres per day. Similar large volumes are needed for crops under glass where as much as 56 000 litres per hectare can be used during a hot sunny day.

Spray irrigation may also be used as a frost protection measure when fruit buds have formed and frosts may still be expected. A layer of ice forms around each bud rather than the moisture in the bud freezing and destroying the crop. The quantities used are high, equivalent to 25 mm of rain each night and there may be a need to spray on 12 or more nights each year, depending on where you live.

The large water requirement is the biggest problem in planning a spray irrigation scheme and it is important to ensure that you have accurately calculated how much water you need. The amount of water used by crops depends on the type of plants involved, the soil type, the season and the amount of rain which has been falling recently. Shallow rooting plants tend to dry out more easily and therefore need more watering. Sandy soils drain easily and tend to dry out more quickly than loams. In the growing season plants tend to have a high water need as a high proportion of their new tissue is made up of water. Temperature is another important factor in determining how much moisture plants need and so this adds another aspect to the seasonal dimension. Clearly the amount of rain which has fallen during the past few days has a direct bearing on the need for irrigation. From all these factors you can see that the calculation of spray irrigation water requirements is very complex and outside the scope of this book. I suggest that you obtain advice on the water needs from the government agricultural department in your country.

FISH FARMING

The importance of a good water supply in this part of the agricultural industry goes without saying. A number of different varieties of fish are grown on fish farms with the main ones including both rainbow and brown trout, salmon and carp. There are two types of fish farm: hatcheries which usually produce their own eggs from brood stock, and rearing farms which buy small fish and grow them to a size suitable for the market they supply. Salmon rearing requires sea-water for the fish to grow to full size but the young fish are hatched and reared in similar circumstances to trout. Most of the fish are for the table but a significant number of fish are used to re-stock rivers and lakes where the natural fish population has been depleted by overfishing, pollution, or even where large-scale water abstraction schemes have affected salmon or trout spawning grounds. Carp are grown both for food and for stocking ornamental ponds, with large colourful fish fetching very high prices.

Only relatively small quantities of water are needed for fish breeding but water quality is of paramount importance. Fish rearing farms, on the other hand, need enormous quantities of pure water which must run continually through the stock ponds and then be discharged to waste. The amount needed is related to the number of fish on the farm and a flow of about 430 000 litres per day is required per tonne of fish. This means that the location of the fish farm, particularly one for fish rearing, is dictated by the presence of an adequate water supply. If you try to site it from any other point of view it may well be economically unviable.

When choosing your supply you should bear in mind the cost of producing the water, as well as the stringent quality requirements referred to above. Further details of these quality requirements are given in Chapter 6. If the supply is to come from a well or borehole and all the water has to be pumped, the economic viability of a fish farm may well be in doubt. The best types of source are springs and streams where gravity flow can be used.

SMALL BUSINESSES

There are many businesses being established in the country which require a private water supply. Hotels, guest houses, restaurants, pubs and caravan and camping sites are appearing in increasing numbers, often in areas where public water supplies are poor or not available. The best way to calculate water requirements for such businesses is to use the figures given in the section on domestic supplies, being careful to make realistic modifications to these quantities. There is no point, for example, in assuming that people on a camping site will use as much water in washing clothes or even pots, as many meals may be taken away from the camp site. On the other hand, if you are running a hotel you will need to take into account water needs in the kitchen and laundry as well as the water used by guests for baths, washing, etc. If you send your dirty sheets and towels to a laundry you will not need quite these quantities and should make an appropriate adjustment. These calculations are basically common sense and simply a matter of sitting down and thinking through all your water needs.

Over the past 10 years or so, the popularity of bottled waters has dramatically increased. A viable bottled water business needs a reliable supply of constantly good quality water, and obviously, the quantities available from a source are paramount in determining the success of the business. Chapter 3 described how to assess the reliable yield of a water source. In thinking about the possibility of using a source for bottling assume that you will only be able to bottle about 50–60% of the total yield to allow for wastage, washing down, cleaning bottles, etc.

WATER SAVING

It seems appropriate to think about the ways to reduce water wastage and reducing overall consumption while we are considering how much water is needed. Leakage from ageing reservoirs and tanks or old water pipes and other fittings can account for staggeringly large water losses. In recent years in the UK, for example, this aspect of water companies' operations has been given a lot of attention in the media. It has been shown that typically water losses can be more than 30% of the total water abstracted from the sources and extreme examples can be well over 50%. The cost of such wastage is high if the water has been pumped and treated. In some examples of private water supply systems pumping may not be necessary and the water only treated as it is being used so you may regard the costs as practically zero. However, leaking water can cause damage to buildings or result in dampness problems. The main disadvantage of leakage, however, may not be cost but simply that there may not be enough water available for your needs.

The best cure for water leakage and loss prevention is to carry out regular maintenance, which is discussed in Chapter 8. Water consumption may also be reduced by water being used more than once. For example, water used for washing may be suitable for watering plants or cleaning the car.

Consumption can also be reduced by using water-saving appliances. We have already discussed the large volumes required for flushing lavatories. It is possible to reduce the amount used by reducing the volume of water stored in the cistern or buying a lavatory with a dual flush. These usually have a valve which controls the amount of water released, reducing it to about half unless the handle is kept depressed. To reduce the volume of water in your lavatory cistern adjust the arm on the float valve so that it switches off at a lower level. You will have to experiment as to how much water you need for the loo to flush properly. No savings will be made if you reduce the volume stored in the cistern so much that repeat flushing is needed to clean the lavatory. An alternative may be to change your water-operated lavatory for one which uses chemicals to treat the sewage or to adopt a system which does not require water, such as those described in Chapter 10.

If you have automatically flushing urinals in your building it is a good idea to look at the available devices for reducing water use. Sensors which detect the presence of each user can be fitted to the system so that flush cycles only operate when needed and avoid flushing during quiet times.

Many modern domestic appliances such as dishwashers and automatic laundry machines are designed to use less water and, in fact, to be altogether more economical in their operation. This is achieved by recycling the water in the machine so that it is used several times in a washing cycle. This will reduce both the water and the electricity used. If you collect the water as it is discharged from the machine instead of letting it go down the drain you will be able to use it to water your vegetables. It is important to check that the detergent used will not cause any harm to you or your crop before you use it in this way.

If hoses are being used it is a good idea to fit a control to the end which will automatically shut off the flow if the hose is put down. The rate that water flows from an open-ended hose is very high and large wastage can quickly result.

5
Building a New Source

There are many different ways in which supplies can be obtained. These range from the simple collection of rain-water to the use of rivers, streams and springs and the construction of reservoirs, wells and boreholes. This chapter describes how these source works are constructed so that you will be able to understand your existing system in sufficient detail to maintain it, or to allow you to construct a new source altogether. When you come to plan the detail of building a new water source think carefully about the quality of the materials you will use, especially if the water is for drinking. It is important to select only those materials which are specifically made for drinking water supplies; other materials may not be manufactured to the same exacting standards and could introduce harmful material into your drinking water.

RAIN-WATER COLLECTION

Rain-water collection or rain harvesting is used today in many countries where alternative sources are not easily available. In countries such as Britain these systems were widely used into the end of the last century as part of water supplies to individual houses. They can still be found in some older houses but their use is very much reduced because of the widespread availability of public water supplies and the fear of pollution of such sources. Rain-water collection systems are used for part of the public water supply in the Isles of Scilly, off the south-west tip of England. Recently, however, a desalination plant has been installed to provide fresh water from sea-water to help meet the increasing demand from the tourist trade.

These collection systems tend to be of two main types. The first was intended to provide a soft water supply and to be a supplement to other sources such as a spring or well. Rain-water is very soft and was collected specifically for washing clothes in times before modern detergents had been developed. In this type of collection system part of the house roof drainage was fed directly into a storage tank in the roof space. From here it would be piped to a separate cold water tap in the kitchen or scullery. The tanks used were often of fairly small capacity, perhaps only 225 litres or so, and so there would be a need for fairly regular rainfall. The main problem was caused by the frequent use of lead-lined wooden tanks, as the soft water dissolves the lead making the water dangerous to drink.

The second type of system is where rain-water collection is the sole source supplying the needs of the household. Here all the drainage from the house roof area and perhaps some outbuildings is fed into an underground storage tank or cistern. The tank has to be of a sufficient capacity to overcome droughts and so the most practical place to locate it is underground. This also has the advantage that the water is kept cool and it will also be dark which inhibits bacterial and algal growth. A typical arrangement for rain-water collection is shown in Figure 5.1. If you are thinking about constructing one of these tanks you will have to have one large enough to withstand a long period without rain. Total lack of rain is not the only problem. The difficulties of the UK drought in 1976 were caused by rainfall being 62% of the average between October 1975 to March 1976 rather than the seemingly endless days without rain during the following summer.

You will need to pump the water from the storage tank into your distribution system. In the old days a hand pump would have been used but now an electric pump is the most convenient. It is essential to make sure that vermin, including insects, cannot get into the tank. Collect the water from the roof with a standard gutter system but pass the water through a small chamber before it goes into the storage tank. This small chamber will act as a silt trap and prevent debris from getting into the main supply.

(c)

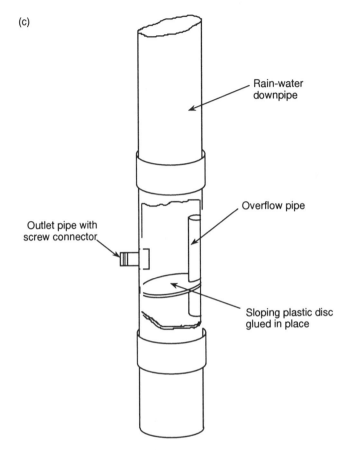

Rain-water
downpipe

Overflow pipe

Outlet pipe with
screw connector

Sloping plastic disc
glued in place

Figure 5.1 If you do not need all the water flowing from your roof you can take part of it by fitting a device to your downpipe like the one shown in (c). There are several different ones on the market or you can make your own. Take care to prevent leaves entering the downpipe and blocking it; this regularly happened to the one I fitted to my house until I fitted a mesh "cage" to the top of the pipe to keep the leaves out

Disinfection

Before using your rain-water collection tank to store water you must disinfect it using a chlorine solution as described in Chapter 6. It is important to make sure that bacteria do not grow in the tank and this will entail disinfection on a regular basis.

Glass-house Irrigation

An important use for rain-water collection systems is to supplement irrigation water in commercial glass-houses. It seems a great waste for rain-water which runs off the roof not to be used when large quantities of water are needed to irrigate the plants inside the glass-house and sheltered from the rain.

Design your tank to take all the rain which falls on the roof to get the maximum benefit. You will need to know the amount of rainfall in your district on a monthly basis as well as the area of your glass-house roof. Take the rate of water use in the glass-house into account so that the capacity of your tank is not much bigger than necessary. Do not make the mistake of trying to use your well to store this water. If you do, the water which runs into the well from the glass-house roof will simply flow out of the well into the aquifer and you will lose it.

DEW AND FOG

The parts of the hydrological cycle represented by dew and fog are not normally thought capable of providing a source of water and cannot be recommended for new development except in the most extreme circumstances.

Dew Ponds

Dew ponds are shallow pan-shaped depressions made to collect water for animals and are a traditional feature of the landscape in the Downlands of southern England. The ponds are 10–40 m in diameter and made from puddled clay or even concrete. The name is misleading as they mainly collect rainfall and a very small amount of local run-off, rather than condensing dew which is thought to make up only about the annual equivalent of 25 mm of rain. They are of limited value in most water supply schemes because they have limited storage and generally dry up frequently.

Fog Harvesting

In those parts of the world where fogs occur frequently it is possible to use them as the basis of a small water supply system. Experiments have been carried out in the Atacama desert in northern Chile to collect water which condenses onto a fine polypropolene mesh to provide a water supply for a village of 350 people. The collection system consists of double layers of mesh-forming panels, each of which measures 12 m by 4 m, hung about 2 m above the ground facing into the wind. In all, 75 panels were used to provide 11 000 litres/day, with the tiny fog droplets running down the mesh and collecting in a trough which runs into a storage tank. It you have no alternative, it may be worth thinking of trying this unusual source for your drinking supply. Daily yields of 3–4 litres/m^2 of mesh have been reported, but success depends on finding the optimum position on the hillside. You will have to experiment by erecting small screens in different locations and comparing the yield. Do not look to obtain large quantities however. The Atacama experiment showed that the water diverted into the water supply system had a devastating effect on the local wild plants and animals which are entirely dependent on the fog as their only source of water. Consequently fog harvesting can only be justified for limited individual supplies where the amounts taken are small.

RIVER AND STREAM INTAKES

There are significant drawbacks in using a direct intake from a river or stream as your sole water supply. First, you can only get a supply if there is water in the stream. Secondly, your supply will be vulnerable to pollution problems. There are a number of common causes of pollution. Animals may drown in the stream and become wedged near your intake or there may be animal droppings which contaminate the supply. Septic tank overflows to the streams and accidental spillages are other frequent problems.

If you are happy that the stream or river you want to use will not dry up and there is no danger of pollution, you can construct an intake fairly easily following the principles set out in Figure 5.2. When installing an intake make sure that the end of the pipe will always be submerged. Avoid silt or other fine material from getting into the end of the pipe by fitting a strainer. The sort of strainer often used resembles the rose on a watering can; alternatively you can use one of the porous pot types which are similar to the filters described in Chapter 6.

Small intakes can be made extremely simple. I have seen one which simply consists of a piece of pipe fitted with a strainer, which

(a)

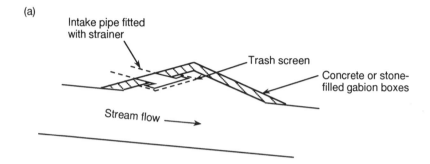

Intake pipe fitted
with strainer

Trash screen

Concrete or stone-
filled gabion boxes

Stream flow ⟶

(b)

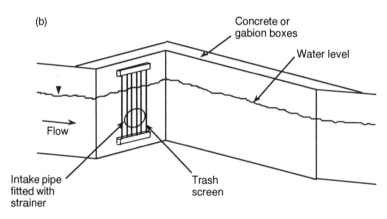

Concrete or
gabion boxes

Water level

Flow

Intake pipe
fitted with
strainer

Trash
screen

Figure 5.2 (a) plan view; (b) side view. Build an intake on a water course so that the flow does not go directly into the pipes as this will encourage floating rubbish to be washed in. Dig a trench for the intake pipe to a depth where it will be submerged all the time. Protect the bank around the pipe with concrete or gabion boxes (wire-mesh) filled with stones to prevent erosion. Fit a strainer on the end of the pipe to prevent small objects entering and a trash screen comprising metal bars to keep the large objects off. During construction you will need to make a temporary dam using sandbags to keep your working area dry. Make a small temporary sump at one end so that you can easily pump out any water which gets in

is buried in the stones in the bed of a stream. Although this system is rather crude it has worked for many years without suffering from many serious problems. A more sophisticated alternative is to connect the pipe to a small tank buried in the stream bed like the example shown in Figure 5.3. The tank is filled with graded gravel or rubble rather like a soakaway and this acts as a filter. If you do not use this type of filter tank you should construct a settlement tank near the intake to prevent sand or silt getting into the supply

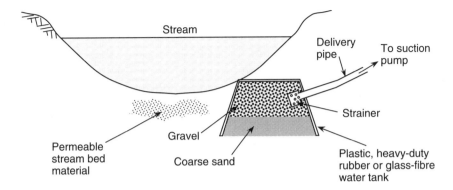

Figure 5.3 An alternative river intake can be built using a tank filled with graded gravel or rubble rather like a soakaway. The sort of plastic tank used in domestic plumbing systems will work very well. It is protected from damage by being completely buried in the stream bed. It must be buried upside down so that the water enters the filter box through a substantial thickness of river bed material. This will give added protection to that provided by the filter material in your box. This type of system will only work where the stream bed is sandy or at least sufficiently permeable to allow water to enter the tank. Make sure that the filter material is clean by disinfecting it if necessary following the procedure described in Chapter 6. Avoid using limestone gravel as it is likely to fuse, reducing the rate of flow through the system

pipe. These settlement tanks are also called sand traps and are described in Chapter 7.

Another way to make sure that stream water levels are always high enough to keep your intake pipe submerged is to install a low weir. This need not be very sophisticated at all. A length of steel pipe about 75 mm or 100 mm in diameter will control the water level. It should be weighted by filling it with stones, clay or concrete and should be firmly staked to the river bed. The stakes should be driven at least 50 cm into the stream bed at 1 m intervals along the pipe. Old scaffolding pipes are ideal for this job. Some countries have legal controls on building weirs and, if you live in England or Wales for example, you may need a licence for this weir (see Chapter 9).

SPRING COLLECTION CHAMBERS

Some springs issue at a single point, whereas others consist of a general seepage which needs collecting together before it can be used for a water supply. Water from both sorts of spring should be collected in a small chamber to protect it from pollution. It is essential that you take proper precautions to prevent surface water from entering the catchpit. Recent tests have shown that intermittent pollution of spring supplies can occur with each flush of water from rainstorms. The collection chamber also provides storage to draw on at times of peak demand. It can also act as a header tank and drive water through the supply pipes under gravity as described in Chapter 7.

Collection chambers can be constructed of a wide variety of materials, including bricks, concrete blocks, stone blocks and concrete rings. A chamber such as the one in Figure 5.4 should be built over the spring so that water flows up into the base of the chamber or in at the side. This will probably mean digging into the hillside round the spring. Do not worry about damaging a spring when you do the digging. The spring is draining an aquifer and is being pushed out by a great head of pressure. You would find it very difficult to stop the spring flowing, even if you wanted to!

Land Drainage Systems

In Chapter 2 a type of spring was described which is really part of an artificial land drain system. Quite often these drains consist of a stone-lined culvert buried at a depth of 1–1.5 m. At various places along the length of the drain there are cattle watering troughs and catchpits used for domestic water supplies. The main problems are contamination from animals using the troughs and collapse of the drain which could allow further pollution of supplies. Maintenance may be difficult where the drain runs through a number of different properties and is used by various owners.

I once came across a land drainage supply which illustrates all of these problems. A catchpit near the top of the system was disused and neglected but all the water had to flow through this tank. It contained dead animals which contaminated the whole supply. Further down the hillside part of the drain had collapsed allowing sheep droppings to enter it. At the bottom of the hill was a collection chamber used for domestic supply to three separate households. Beware of accepting cups of tea in such circumstances! If you have a water supply based on this type of system I suggest that you either replace it with a borehole or take extensive measures to protect it from pollution.

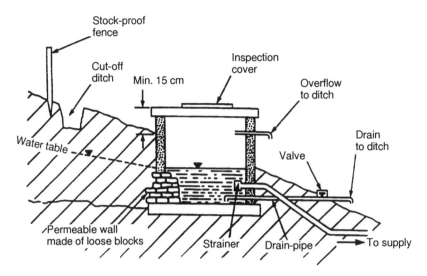

Figure 5.4 Build a chamber round a spring to protect it from pollution and to provide storage. First dig into the hillside to expose the spring. To build a chamber you will need to make a concrete slab for the foundation. This is likely to mean diverting the spring flow before you carry out the work. Use sandbags, pipes and polythene sheeting to make this temporary diversion. If necessary, dig a temporary trench to carry the flow but make sure that you properly back-fill and seal it once the job is finished. If you are using wet concrete you will need to allow several days for it to cure before building the chamber itself. Build a wall of loose blocks at the back of the chamber for the water to flow through. The walls can be made from concrete blocks, bricks or reinforced concrete. Alternatively you could use a modified plastic water tank, protected by a masonry wall. The top of the chamber must be at least 150 mm above the ground to prevent surface water from getting in and it should be fitted with a lockable watertight inspection cover, large enough to allow access for maintenance. The water supply is taken from the chamber through a pipe installed above the base to avoid it being covered by silt and fitted with a strainer. An overflow will be needed to get rid of excess water at times of high flows. The overflow pipe must be large enough to take the maximum flows and this is one reason why you need to measure the spring before building the source works. Again, take care to ensure that your collection chamber is vermin-proof as you will not want a dead mouse or frog in your drinking water. It is important to make sure that the level of the overflow pipe is not so high that the normal spring flow cannot get out of it. If the head of water in your collection chamber builds up so that it is greater than that in the aquifer near the spring, it may cause water to break out at a different place on the hillside, causing your supply to fail. The sanitary arrangements for your spring should include a stock-proof fence to ensure that farm animals do not get within 8–10 m of the chamber. Dig a ditch above the spring to intercept surface run-off which could contaminate your supply. A new source should be disinfected with a solution of sodium hypochlorite or bleach before the water is used. More information is given in Chapter 6

A further problem with water supplies from these land drains is their unreliability. In some drainage systems most of the flow is surface water rather than groundwater and flows dwindle in dry weather causing water shortages. It may be possible to overcome this problem by installing a large storage tank but this could have a number of drawbacks, especially for the quality aspects of the supply. This is particularly true if water has to be stored for long periods of time and an alternative source may be a better bet.

GROUNDWATER COLLECTION

On the face of it, this type of source is the same as the land drains described above. Both consist of land drainage pipes of one sort or another feeding into a catchpit. Land drainage systems, however, are primarily designed to drain other fields. Groundwater collection systems, on the other hand, are designed to tap shallow aquifers, such as sands and gravels, and yield more water than wells.

A groundwater collection system consists of a series of land drainage pipes buried in trenches below the water table. These are often in a herringbone pattern typical of tile land drainage schemes. The pipes used should be 100 mm or 150 mm in diameter and can be made of unglazed clay tile drains or porous concrete pipes. More modern land drains are made using slotted plastic pipe which comes on long reels. Continuous lengths of pipe are "ploughed" into the ground using special machines. This will save you a lot of digging but if you decide to use plastic pipe take care to ensure that it is made from suitable material for drinking water supplies by checking with the manufacturer. The drain pipes feed into a collection chamber, which closely resembles a dug well, from where water is pumped into the supply. Indeed, the yield of dug wells can be improved by linking them to this type of collection pipe. There are no set rules for positioning the pipe runs, except to make sure that they generally follow the direction of natural drainage. This often follows the slope of the land.

Unfortunately, these collectors are subject to the same pollution risks as are all shallow groundwater sources. Furthermore, their yield will deteriorate with a lowering of the water table. This problem can be kept to a minimum by careful construction. If you live in the northern hemisphere the best time to dig the trenches and install the pipes is September or October when groundwater can be expected to be at its lowest. For countries south of the equator the best time is during March and April. Make sure that the pipes are buried as deep as possible below the water table and that they slope towards the main collection chamber. Install a few inspection chambers along the line of the pipes so that you can rod them if they become blocked.

This type of source is ideal for applications like fish farms where they can be expected to give the high yields needed and provide water at a constant temperature. If the collector is constructed in river gravels near a stream or river, its yield will be maintained by the river water. You will enjoy the advantages of a reliable supply and avoid most of the quality problems associated with river abstractions.

Groundwater can also be collected by excavating a large lagoon or seepage reservoir in shallow aquifers where the water table is near the surface. Such a reservoir will provide the large volume of water needed for an irrigation project, for example. Yields are high as pumping can be at a greater rate than would be possible from wells in the same aquifer.

Such schemes have several drawbacks. The provision of a large area of open water will significantly increase evaporation losses. This will reduce the volume of water available for supplies. The large pond you have built will soon be full of plants and weeds which will alter the quality of the water. This may not necessarily be a problem in itself but a by-product of the processes may be the deposition of slime in the bottom of the lagoon. This can reduce the inflow of water when you are pumping. These lagoons will require regular maintenance if they are to be used as a source of water in an irrigation project. However, they have been used successfully to provide ornamental ponds, fisheries or a suitable habitat for wild-fowl.

RESERVOIRS

In this context the term reservoir is used to mean a construction which will hold a volume of water. Reservoirs are used as a source of water for such things as spray irrigation, frost protection for soft fruit blossom, fish farming and industrial processes. In some cases, a reservoir may be wanted to provide a fishing lake or simply for its ornamental effect. The only type of construction which can be recommended for these small reservoirs from both a cost and safety point of view, is the use of earth embankments and dams to hold back the water. Such reservoirs come in two general categories: off-stream and impounding reservoirs.

Off-stream Reservoirs

Off-stream reservoirs can be sited on any suitable land and are fed with water taken from a stream. Ideally, this water should flow to the reservoir but if this is not possible, the water will have to be pumped. These reservoirs should be constructed on impermeable soil to prevent water loss. Artificial waterproofing techniques, such as butyl rubber sheet lining, can be used, but these are very expensive and should be avoided if at all possible.

A very important factor which usually controls the cost of construction is the ratio between the volume of earth dug out to make the reservoir and the volume of water which can be stored in it. If the water level in the finished reservoir is to be near original ground level, the ratio of water to earth will be less than 1:1. Such a reservoir would be uneconomic to construct. The usual way of ensuring that the maximum volume of water can be stored in an off-stream reservoir is to use the earth from the excavation to build banks. This increases the storage volume because the final water level is above the natural ground level. The embankments must be made of suitable material and constructed in a way which will ensure that they are safe and will not leak. The main features of off-stream reservoirs are shown in Figure 5.5.

Impounding Reservoirs

The second type of reservoir is where a dam is built across a stream to impound water behind it. Once again, it is very important that the correct site is chosen so that leakage will not occur through the ground around the dam. There must be adequate room for a spillway which will allow flood waters to bypass the dam and not wash it away. Ideally, a site should be chosen so that the dam can be

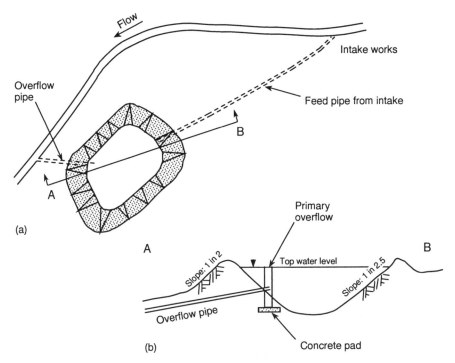

Figure 5.5 (a) plan view; (b) cross-section. An off-stream reservoir is built by digging a hole and using the excavated material to build the embankments. The reservoir will leak unless the soil has a high clay content or it is lined with a suitable sheeting. Butyl sheeting is most frequently used but is very expensive and could make a reservoir scheme uneconomic. Before making the embankments remove all the top soil as this could provide a leakage path. The soil should be saved and spread on the outside of the embankments which are sown with grass to help provide stability. Build up the embankment in thin layers no more than 250 mm at a time. The material should be consolidated using a tractor for example, which runs over each layer after it has been placed. The embankment slopes should be 1 in 2.5 on the inside and 1 in 2 on the outside. Fill the reservoir by constructing a river intake upstream of the site so that the water can flow in under gravity to avoid pumping costs. Place an area of concrete at the inflow end of the pipe to prevent erosion and on large reservoirs it will be necessary to put stones round the edge at the water line to prevent erosion from wave action. Keep the water level controlled by building a primary overflow similar to the one shown in Figure 5.7

built using material excavated from the reservoir site. Bringing material in will mean that construction costs will be increased and the dam may be uneconomic as a commercial proposition.

Seepage from the site may be difficult to overcome. This can happen if permeable rocks such as sand or gravel occur on the reservoir floor or in the valley sides. An impounding reservoir will occupy a natural valley and take up a larger area of land than an off-stream reservoir of the same capacity. Lining the latter is expensive enough, but if you will have to use imported clay or butyl rubber sheeting to prevent leakage in the case of an impounding reservoir, it is probably advisable to abandon your plans for that site.

Spillways

Any impounding reservoir must allow all natural stream flows to get past it without causing structural damage. This is usually achieved by constructing two spillways: a primary spillway to take the usual daily flow and a storm water spillway which will take flood flows. It is important that both spillways are large enough to cope with the flows you might expect in the stream which has been dammed.

Quite often the primary spillway consists of a chamber constructed to the final level of water in the reservoir down which the water cascades and flows beneath the dam through a pipe. These spillways can be constructed in masonry or concrete rings. The pipe which goes beneath the dam should be constructed before the dam is made and precautions taken to prevent leakage along the outside of the pipe.

It is important that storm water spillways are excavated in solid ground and this is usually done to one side of the earth embankment. Every care should be taken to make sure that the spillway is large enough to cope with the maximum flood flows you can expect in the stream catchment. It is better to have allowed for greater floods than will occur than to have your dam washed away in a catastrophic flood because you have not built the spillway wide enough.

Site Investigation and Construction

Before choosing a site for either sort of reservoir you must find out if the underlying rock is permeable. This is best done by digging various trial holes across the site so that samples of rock and soil can be examined. If you are feeling fit you could dig them by hand but usually it is better to use a small back-acting excavator. These machines can dig holes to depths of 3–5 m. Trial holes will give enough information about the watertightness of the site and the suitability of site materials for the construction of the embankments, but you may well need to employ an expert to interpret this information.

In constructing an earth embankment to retain water, either as part of an off-stream reservoir or as the dam across a stream, you must ensure that the embankment is structurally stable under all conditions and is watertight. The stability of such low embankments depends upon the type of soil in the foundations. The type of material used in the construction of the embankment and the overall width and slopes of the sides of the embankment will also affect its stability.

Dam foundations must be strong enough to support the weight of the embankment without undue settlement. Most natural materials have sufficient strength to bear the weight of the sort of low embankment considered here, but topsoil, peat and any soil which contains organic matter is unsuitable and must be stripped off and removed from the site. It can be retained and used as a final cover for the embankments so that they can be grassed.

The stability of the embankment itself is ensured by keeping slopes to 1:2 on the downstream side and no more than 1:2.5 on the upstream side. All material should be placed in layers of about half a metre at a time. If you are going to include a clay core in the dam this material should be placed in layers no more than 250 mm in thickness. The top of the embankment should be almost flat but cambered or with a slight fall to prevent puddles forming during rain. It should be no less than 2.5 m wide if the embankment is 2 m or less in height. If the embankment is as high as 5 m the top should be increased to 3 m wide.

Once the embankments have been made they should be covered with the soil you have stripped from the flooded area on the top and downstream sides. The wet slopes should be protected against wave action by a layer of stone. The soiled areas should be sown with grass which will hold the surface together with their roots. It is important to prevent trees or bushes from growing on an embankment as sooner or later water will "pipe" through the embankment along the lines of the roots.

These embankments can be constructed using the sort of road-making machines which can be hired from a plant hire contractor.

You will not be expected to drive these machines yourself as the contractor will provide an experienced operator. If you carefully mark out where you want soil stripped and stored, and where the embankment is to be built, the operator should be sufficiently experienced to make a good job of it. Make sure that the material is put down in layers and compacted using the weight of the machine. Further compaction will probably be needed. This can be achieved by driving a tractor up and down along the centre line of the dam

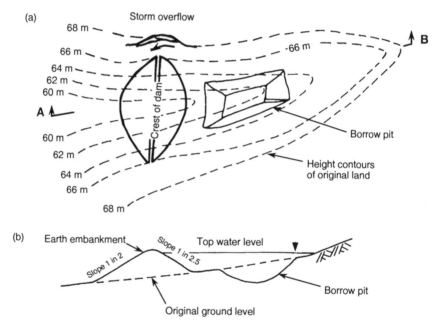

Figure 5.6   (a) plan view; (b) cross-section. A small earth dam can be built across a valley to create an impounding reservoir. It is important to choose a location where the shape of the ground means that only a relatively short embankment will be needed and where the geological conditions will not permit leakage. It is essential to have an accurate survey of the site so that the dam can be designed. Scrape off all the top soil from the area of the dam and the area which is to be flooded. Again this can be spread over the outside of the embankment and sown with grass. The material for the embankment is obtained from a borrow pit inside the area to be flooded which will also increase the storage capacity of the reservoir. Again build up the embankment in layers, consolidating it as you go. The inside slope should be 1 in 2.5 and that of the outside 1 in 2. As the dam will impede all the flow of the stream it is important to take steps to prevent floods from washing it away. To achieve this, cut a storm overflow channel at one end of the dam in the solid ground. It is important not to build this channel on the embankment you have constructed as it will be eroded during flood flows. The storm overflow channel should slope gently up from the water edges to a high point some 0.7 m below the crest of the dam. From there it should slope to the stream channel at a slope of 1 in 40

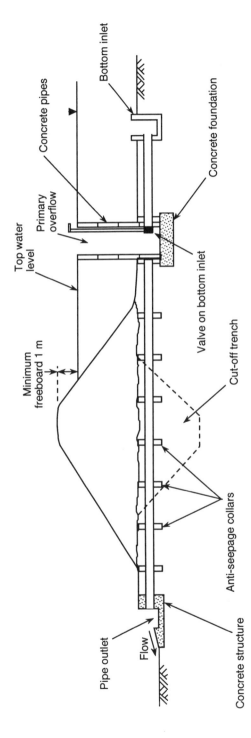

Figure 5.7 Where the ground is not sufficiently clayey to prevent any leakage, dig a cut-off trench beneath the dam. It should be back-filled with suitable material in layers just as the dam is built. It may be necessary to import clay for the job. In this case you do not need to build the whole of the embankment in clay. Make a strip of clay at least 2 m wide, built up in layers and consolidated as before. The rest of the embankment can be made from the material obtained from the site. Build a primary overflow to take the normal excess inflow. One way of doing this is shown here, where a chimney is built up of concrete pipes set on a concrete foundation. The top of the chimney is set at the top water level for the reservoir and at least 1 m below the top of the dam. So that you can entirely drain the reservoir if necessary, you will need a bottom inlet. This should be connected to the primary overflow as shown and controlled by a valve accessed from the top of the primary overflow. The water which flows through the primary overflow is conveyed below the dam through pipes set in a trench dug into the undisturbed ground. A series of anti-seepage collars should be built round the pipe. These are made by digging a short trench at right-angles to the main trench and filling them with concrete. The collars should have a diameter five times that of the pipe and be set at intervals of twice that distance. Build a concrete outlet bay where the pipe discharges into the natural water course to prevent bed erosion

during construction. Adequate compaction is needed to prevent leakage.

Figures 5.6 and 5.7 show some of the general principles of earth dam construction. Detailed consideration of the design and construction of these reservoirs is really beyond the scope of this book and you should consider employing a suitably qualified engineer to design and supervise the construction of your reservoir.

WELLS AND BOREHOLES

Both these terms are in everyday use to refer to two different types of water well. A well is generally taken to mean a shallow large-diameter excavation, usually round in plan and perhaps constructed by hand, although these days they are often dug by a back-acting excavator. The term borehole is usually taken to embrace all deep small diameter wells which have been excavated by cable tool or rotary drilling methods.

Wells

The traditional image of a well is a circular brick- or stone-lined hole about a metre in diameter with the lining material extending a metre or so above ground level. On top is a picturesque roof which perhaps is thatched or tiled. Water is abstracted by means of a bucket attached to winding gear suspended between the roof supports.

This picture of a wishing well is not so far away from the truth as you may imagine. In fact, such a well would have all the essential features needed to give a good and pure supply of water. The well needs to be lined to prevent it collapsing and the lining must extend above ground level so that surface water cannot contaminate it. There needs to be some sort of cover to prevent dirt from getting in, but perhaps one more substantial than a roof. There also needs to be some means of abstracting the water, although you will probably find that an electric pump is much more convenient than a bucket on a rope.

The traditional method of constructing a well was to dig a hole about a metre deep and wide enough to lay a cart-wheel in it. In some cases an old wheel may have been used, but generally a purpose-made iron-shod curb was placed in the hole and used as the footings for a dry brick or stone wall. This wall is termed the *steening* from an old English word for stone. The inside of the hole was dug out, periodically undercutting the wheel so that the weight of bricks or stones pushed it down. The lining was then built up to ground level again and then more material dug out of the bottom of the hole. In this way the hole was dug out until the water table was reached. Digging continued but pumps were needed to lower the water level so that the well could be dug as deep as possible. Once digging was completed the joints in the brickwork above the water table were mortared as a precaution against surface water entering the well.

The limited space inside the well usually restricts the number of men digging to one at a time, so well construction by hand is a slow business. The hard work means that the well digger needs a good air supply. The carbon dioxide which we breath out is heavier than oxygen and so tends to accumulate in the well. It is very important to monitor the oxygen content when working in a well and if necessary use a fan and hose to blow fresh air down the hole.

There are no reasons why these methods cannot work today provided you take proper safety precautions when working in a hole deeper than 1.5 m. Always make sure that the lining material is adequate to prevent the collapse of the well and check the ventilation. Never work alone. Always have someone at the surface who can help you out or raise the alarm in case of an emergency. The most likely accident is to be hit on the head by the bucket which is used to haul out excavated material so take care and wear a safety hat.

There is no need to scour the countryside for an old cart-wheel as it is possible to make a better cutting shoe out of concrete, as shown in Figure 5.8. Make sure that the concrete has set and cured before you start to dig the well. Take out an equal amount beneath the cutting shoe so that it moves down evenly, keeping it as horizontal as possible. This is not likely to be a problem once you have built up a metre or so of brickwork on it which will provide the weight to push it down. Variations in soil conditions may cause a problem so watch out for any sign of change and take it slowly through these more difficult parts.

These days it is usual to line the well with large-diameter concrete pipes of the sort normally used for manholes. These pipes generally have a type of joint (known as an ogee) that is designed to lock the

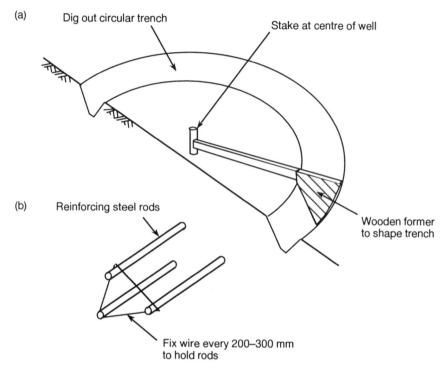

(a)   Dig out circular trench

Stake at centre of well

(b)   Reinforcing steel rods

Wooden former
to shape trench

Fix wire every 200–300 mm
to hold rods

Figure 5.8   A cutting shoe for digging a well can be made from concrete. Dig out a mould in the position where you want to dig the well and you will avoid the problems of having to move it (a). To ensure that the mould is the correct shape, make up a piece of wood to the exact cross-sectional shape of the cutting shoe. Then attach it to a piece of wood as long as the radius of the well. By placing the end of this pole in the centre of the mould you can rotate the shaped piece of wood to make sure that the mould is symmetrical and the correct shape. Make three rings of reinforcing steel rods and secure them in place with wire (b). Use a normal 1:2:4 strength concrete and allow from four to seven days for it to cure before digging the well

pipes together when they are used in a vertical position. When choosing the size of pipes you are to use, bear in mind their weight and consider how you are to put them in position. As far as I am aware, it is not possible to buy porous concrete pipes above 250 mm in diameter, but large-diameter pipes are available with holes made in them. You will need to have several lengths of these so that water can get into the well.

You do not need to dig the well by hand either! Hire a hydraulic back-acting excavator (as shown in Figure 5.9), which will be able to dig a hole to a 5 m depth in less than half an hour. Once the pipes have been installed it may be necessary to deepen the well by hand

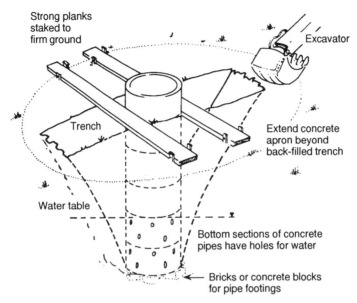

Strong planks staked to firm ground

Excavator

Trench

Extend concrete apron beyond back-filled trench

Water table

Bottom sections of concrete pipes have holes for water

Bricks or concrete blocks for pipe footings

Figure 5.9   There is no need to dig a well by hand. Hire a hydraulic back-acting excavator which will be able to dig a hole to a 5 m depth in less than half an hour. It is possible to lower the concrete pipes into the excavation using the excavator arm provided that the weight of the pipes is within the safe working load limit of the machine. You must use a sufficiently strong rope, or preferably a chain, to suspend the concrete pipe. It goes almost without saying that the perforated pipes go in first, with the ordinary ones placed on top to bring the well lining to ground level. Once you have positioned the pipes up to ground level you can extend the depth of your well by digging down inside the pipes by hand. Take all the safety precautions set out in the text and also in Appendix 2 and do not work on your own. Make sure that the pipes are vertical, that they always extend to the ground surface so that loose material cannot fall in on top of you, and that the pipes are supported to prevent them slipping.

The hole taken out by the excavator will be a short trench with the dimensions being dictated by the working space needed by the machine. This will mean that it is necessary to give some support to the top section of concrete pipes and have a safe walk way from firm ground to the well top. Scaffolding planks may provide both support to the pipes and a walkway, but they need to be strong enough to take your weight, supported at each end and secured to stakes driven into the ground so that they will not move. These hints cannot cover every situation which may arise so, before you go climbing into holes, remember that safety is your responsibility. Never get into an unsupported excavation which is more than 1.5 m deep. Always think before you act!

but much of the hard work will have been done by the excavator. It is vital that you take adequate safety precautions before you go leaping around your newly dug well. Safety is YOUR responsibility. Always think before you act! Appendix 2 gives further information on safe working practices.

Even if you do not need to deepen the well beyond the depth excavated by the machine, you will still have to get inside it to make sure that the concrete pipes are seated on firm foundations. Place several bricks with regular spacing underneath the bottom pipe. If the well is dug into sand, put a layer of coarse gravel in the bottom, about 30 cm deep. This will prevent sand from being constantly sucked into your pump when the well is in use. Constant erosion in this way will not only wear the pump out quickly but it will also affect the stability of ground round the well and one day it may collapse.

The next step is to make sure that the excavation on the outside of the concrete pipes is back-filled to prevent surface water from entering the well. You need to seal the joints in the upper part of the lining and prevent water movement down the outside of the lining material. Back-fill the perforated part to the top of the water table using coarse sand or gravel. Try to get a siliceous gravel, such as quartz or flints. Avoid limestone chippings which can gradually fuse together and so restrict the flow of water into the well. The upper part of the well lining should be made impermeable so that surface water cannot get into the well. The space between the concrete pipes and the sides of the excavation should be filled with weak concrete or grout which should extend at least 2 m below ground level.

It is not easy to completely fill narrow spaces with grout or any other material for that matter. If you are faced with a narrow, deep annular space to fill, the best way of tackling it is to use a grout pump which can be hired from a plant hire contractor but is not available from most DIY shops. Attach pipes to the pump and extend them to the bottom of the space. As the space is filled, these pipes are gradually pulled back. If you do not go to this trouble the material will "bridge" and the lower section of the space will be completely free from grout.

The grout should be a straight cement–water mix with no aggregate. It is important to keep the quantity of water used to an absolute minimum; otherwise as the grout hardens it will shrink, leaving a passage for pollutants to get into the water supply. The ideal grout mix for use in both wells and boreholes is five parts of cement to three parts water *by weight*. This means mixing a 50 kg bag of cement with 30 litres of water. In non-metric units this is a one hundredweight bag of cement with six and a half UK gallons or eight US gallons. The main features of a completed well are shown in Figure 5.10.

Once the outside of the well has been back-filled with concrete to ground level, a large concrete apron should be constructed around the top of the well. Make sure that it slopes away from the well top to keep surface water out. I have seen several examples of wells constructed with the concrete surround sloping inwards when usually a convenient hole is left to drain any puddles into the well. In one instance this concrete area was in use as a stock holding pen — no wonder the farmer's family had frequent diarrhoea!

The potential sources of pollutant which may affect the well are septic tanks, privies, sewer pipes, cess pools and animal manure from stock yards and manure heaps. It is recommended that wells are sited at least 30 m away from such potential pollution sources. Remember always to site your well on the up-hill side of these pollution sources.

Disinfection

After constructing your well it is essential to disinfect it before using the water. Use a strong solution of sodium hypochlorite made up to the manufacturer's recommendation or a solution of bleaching powder (see Chapter 6). Scrub the sides of the well with the solution and pour it into the well water. Allow it to stand overnight and then pump it to waste until you can no longer smell chlorine. It will be necessary to pump out two or three times the volume of water held in the well before the smell clears. This may take a few hours or even more than a day, but it is quite easy to work it out if you know the capacity of your pump.

Boreholes

These days, the most usual solution to a private water supply problem is a new borehole. They are usually successful for small supplies and are quite easy to construct, although it is necessary to employ a specialist contractor except for shallow small-diameter wells. These tube wells, as they are often called, can be made with a

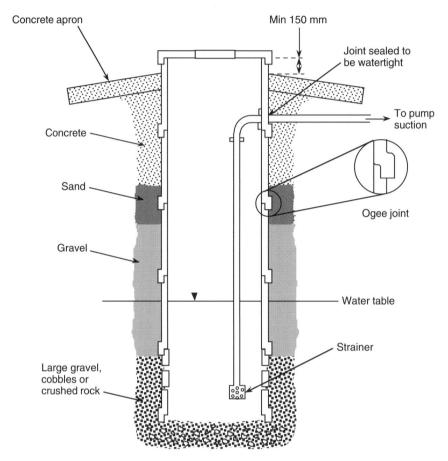

Figure 5.10   This diagram shows the main features of a dug well which is lined with concrete pipes. These pipes have a type of socket joint (known as an ogee joint) which helps to lock the pipes together. The bottom of the well has a layer of large gravel, cobbles or crushed rock to provide a foundation for the rings. Above this coarse material is a layer of gravel which extends above the water table to help water flow into the well. A layer of sand about 300 mm thick is placed on top of the gravel layer to prevent any concrete invading the gravel. The upper section of the well is sealed with concrete which completely fills the annular space between the casing and the ground. The top of the well is at least 150 mm above the finished ground level and has a lid with a lockable inspection cover. The well top should be surrounded with a concrete apron which slopes away from the well and extends a good distance beyond the edge of the ground disturbed during the well construction

hand auger or, alternatively, can be driven into the ground or washed in by a method known as "jetting". Figures 5.11, 5.12 and 5.13 show how you can construct your own tube well using one of these methods. Each method has its advantages and disadvantages. Wells bored by an auger can only penetrate geological materials

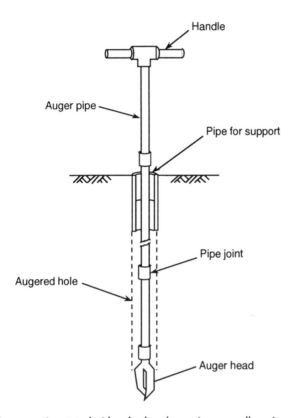

Figure 5.11 An auger is rotated either by hand or using a small engine and effectively screwed into the ground. If the soil is dry or sufficiently clayey the hole will stand for a while without support. Use plastic pipes or tile drain pipes to provide support, placing the pipes in from the surface. Auguring is difficult in saturated sand and loose gravel and it is difficult to cope with cobbles and boulders. Some of the stability problems in saturated sands can be overcome by keeping the hole topped up with water. The method works best in clay soils and once difficult drilling conditions have been reached it is often best to switch to another method to continue constructing the tube well. Bored wells are easily contaminated so their use for drinking water supplies should be avoided. This system can be used to construct wells ranging from 50 mm to 300 mm in diameter, although hand auguring is usually limited to about 100 mm. Depths of 5 m or so can be achieved by hand, and 40 m using a powered tool. The drawing shows a hand auger which is turned by a wooden handle inserted in a holder at the top of the pipes. The auger head is attached to the bottom of a series of socket-jointed steel pipes about 25 mm in diameter

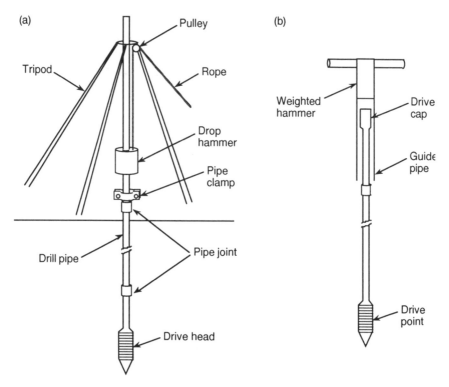

Figure 5.12   Driven tube wells have been constructed for many years all over the world. A weighted hammer is used to hit the top of a pipe thereby driving it into the ground. A special driving point is needed which is strong enough to withstand being driven and is perforated with small slots to allow water to flow in. It may be difficult to keep the pipe vertical but this can be overcome if the well is started with a hand auger. Take care not to pound too hard so that the well point is damaged. If you use an over-size tube as a guide for the hammer it will prevent glancing blows which are likely to cause damage. Turn the pipe periodically with a wrench to ensure that the screw joints are kept tight. This method is the easiest, fastest, cheapest and simplest method of well installation provided that the geological conditions are right. It is possible to drive pipes of 40 mm to 300 mm diameter and to reach depths of 10 m by hand or 20 m if you use heavier mechanical equipment. It works well in sands and gravels but it is not possible to drive through cobbles, boulders or hard rocks such as cemented sandstone and limestone. Too much clay can also cause problems in that the driving head may become too blocked to produce water. If the yield is not sufficient there are several options, including drilling deeper, sinking more wells in the area, trying elsewhere or trying a coarser screen on the drive head. Drawing (a) shows an arrangement using a small tripod to hoist the drop hammer which slides along the projecting drill pipe. A pipe clamp is bolted to the drill pipe above a pipe joint to take the force of the hammer blows. The hammer is raised by means of a rope which runs through a pulley at the top of the rig. Drawing (b) shows a hand-held version with a guide pipe on the bottom of the hammer. The top pipe joint is protected by a drive cap which has been screwed into it

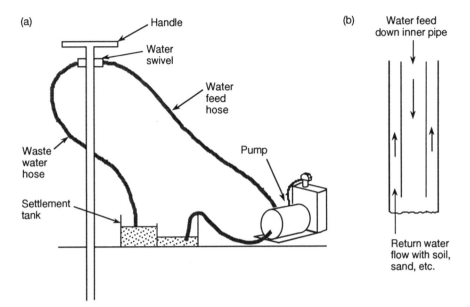

(a) Handle
Water swivel
Water feed hose
Waste water hose
Settlement tank
Pump

(b) Water feed down inner pipe
Return water flow with soil, sand, etc.

Figure 5.13 This method of sinking tube wells is to use water pressure to wash the drive head into the ground. A continuous flow of water under pressure is provided by a pump and directed through a swivel into the top of the drill pipe (a). The water washes the soil, sand, etc., at the base of the hole up through the space between the water injection pipe and the main drill pipe (b). The return water is settled in a tank from which the pump is extracting the water supply. Top up the water supply in the tank if necessary. In some circumstances it is possible to drill a successful well without a second pipe in which case the jetted pipe is used as the permanent support for the well. Otherwise the jetting pipes are removed and the well pipe installed. This has to be done quickly once the jetting head has reached the required depth. Jetting often only takes a few minutes, so have everything prepared beforehand. This method is capable of installing pipes of 50–300 mm diameter to depths of 15–20 m, although 100 mm is reckoned to be the optimum diameter. The method only works in unconsolidated materials and can be used in conjunction with driving or auguring techniques

which contain enough clay to support them during boring. They are generally difficult to seal and are liable to pollution and therefore are not recommended for domestic water supplies. Jetted wells are relatively easy to install and can be sealed to prevent pollution problems. Driven wells cannot penetrate through hard rocks such as cemented sandstone or limestone and are stopped by cobbles or boulders. In the right conditions, however, they can be the easiest, quickest and simplest type of well to install.

When choosing a drilling contractor it is important to find one with the right sort of experience. Many contractors specialise in drilling holes to provide information so that the foundations of roads and buildings can be designed. Others may spend most of the time drilling exploratory boreholes for the quarrying or mining industries. Drilling for these purposes is different to constructing water wells and such companies are likely to have little or no practical experience of water well work. The argument has been put forward many times that "a borehole is a borehole is a borehole", but this is an unrealistic simplification. I would never employ a drilling contractor to construct a well if he (or she) did not have a proven track record in this field.

Before engaging a contractor you should ask for references from two or three of his former clients and then go round and see them. Have a look at the sort of work this contractor does and find out if his old customers are happy with it. It is quite likely that he has worked in your locality before so it should not involve too much effort to go and talk to these people. It is usually very worthwhile to obtain prices from several contractors before selecting the one to do your work. The highest price can be twice that of the lowest. Take care that the offer each contractor makes is on the same basis, otherwise you will not be able to compare one with another and you could end up by employing the most expensive without realising it. In many countries water well drillers belong to a national association which monitors the way their members work. Your local library, council offices or water supply people are likely to be able to give you the association's address.

In some countries the usual arrangement is to pay the contractor an agreed sum only after he has found water — if the borehole fails he does not get paid. Few contractors I know are prepared to work on this basis but you may be tempted to try and persuade them. On the face of it, you seem to be getting the best deal if you do not have to pay for a dry hole. As most boreholes are successful, however, you will probably end up paying more than you need. No one is really prepared to work for nothing and a "no water — no pay" contractor will either increase his prices to cover time wasted on dry holes or cut corners which will give you problems later on.

If you know what you want, you can specify the depth and diameter to be drilled and also the length and type of material to be used for lining tubes. If you are not sure what you want but want to be able to compare prices charged by different contractors, assume that the borehole is to be 150 mm in diameter and 45 m or so deep. A well of these dimensions is likely to provide an adequate supply for a house or small farm. If the water supply is to come from solid rock assume that the lining will be a proprietary screw-jointed well casing to a depth of 15 m. The casing must be sealed with concrete and the borehole top surrounded by a concrete pad at the surface. Remember that if the borehole is to be drilled into sand or a similar material, it will need to be supported for its full depth. You will have to modify your generalised specification by assuming that the lower 15–25 m of casing is the sort with small slots in it to allow the water to flow into the borehole.

If the dimensions of the completed borehole are not quite to the specification it does not matter as a fair price can be agreed with the contractor based on his original quotation.

Contracts

It is best to have a written agreement with your drilling contractor which will form a legally binding contract between you. Make sure that both sign it and each have a copy. The agreement can be simple and written in a straightforward way. It must cover how and when you will pay the contractor and what he will do for the money.

If the final depth is uncertain it is best to pay for the drilling and casing on a "per metre" basis with perhaps a lump sum for him to bring his rig to your site. The agreement should specify the maximum depth to which the rates apply with perhaps an increased rate once a certain depth is reached. In areas where they have worked before, most drilling contractors are likely to have a good idea about the required depth and a sum can be agreed in advance for the whole job including providing and installing the pump. In fact, good contractors will provide all you need including the tank, so all you need to do is connect up the plumbing.

It is a good idea to agree details, such as grouting in the casing and other measures, to protect the borehole from pollution. This can be done using a simple drawing attached to the agreement. Do not be frightened to enter into an agreement. It is a fair way of protecting your interests as much as the contractor's and they should do a better job.

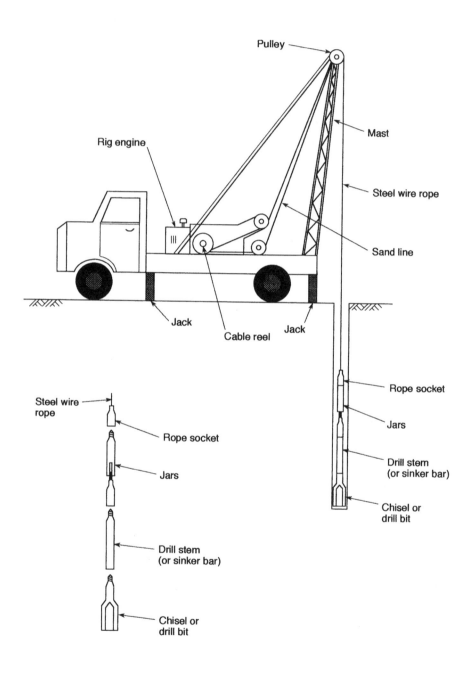

Borehole Construction

There are two basic methods contractors use to construct boreholes. These are *cable tool* or *percussion drilling* and *rotary drilling*. There are several variations of each method but I will only discuss the ones which are in common use. Each drilling method is suited to a particular type of rock. Percussion drilling is usually more effective in soft ground such as sand, gravel, clays and chalk. Rotary drilling, on the other hand, is more successful in harder rocks like sandstone, limestone and granite. Where there is a change in geological conditions with depth both methods may be used to drill a single borehole, although this is less usual when a borehole is being drilled for a relatively small water supply.

Figure 5.14 shows the equipment used in percussion drilling, and photographs of working rigs are shown in Figure 5.15. The main advantages of percussion drilling are that it is straightforward, relatively cheap and is capable of drilling a borehole under almost any conditions. The main drawback is slow progress, which is caused by a need to install support casing as drilling proceeds in sands and other materials which may easily collapse. Secondly, percussion drilling is notoriously slow in hard rocks and in these circumstances it would be most unwise to agree to pay for your borehole on a time rate basis.

Figure 5.14 When using the percussion drilling method, tools are suspended from the rig's mast on a wire rope and continually lifted and dropped down the hole to provide a drilling action at the bottom of the hole. The string of tools (see inset) consists of a rope socket which is a swivel and allows a rotary movement, a set of jars, a drill stem and a drill bit. The most important tool is the drill bit which does the actual cutting. Drill bits look like chisels, are between 1 and 3 m long and can weigh as much as 1300 kg. The drill stem simply consists of a long steel bar and is used to add weight and length to the drill bit so that it will cut rapidly and vertically. The jars consist of a pair of steel bars linked together in the same way as the links in a chain. They play no part in the actual drilling action, but are there to help free the tools should they become stuck in the hole. Under normal tension on the drilling line the jars remain fully extended. If the tools become stuck the line is slackened which allows the links to close together. An upstroke on the line will deal an upward blow to the string of tools. This type of movement is more successful in freeing the tools than a steady pull on the rope, which may only break it. The rock fragments and cuttings are removed from the borehole by a bailer. There are several kinds of bailer but, basically, it is a length of pipe which often has a valve at the bottom. There needs to be water in the borehole for it to work so, if the water table has not been reached, water needs to be added to the borehole. To clean the borehole, a bailer is lowered down the hole and then moved up and down. If a bailer has a valve it fills from the bottom and is more successful in removing debris than one which fills from the top

Figure 5.15 (a) A large lorry-mounted percussion rig working on a drilling project in Ethiopia. The rig has a rotary attachment for use in hard rocks. (b) A small percussive rig being used to construct tube wells in Kenya. (Photographs courtesy of DANDO Drilling International Ltd, Littlehampton, West Sussex)

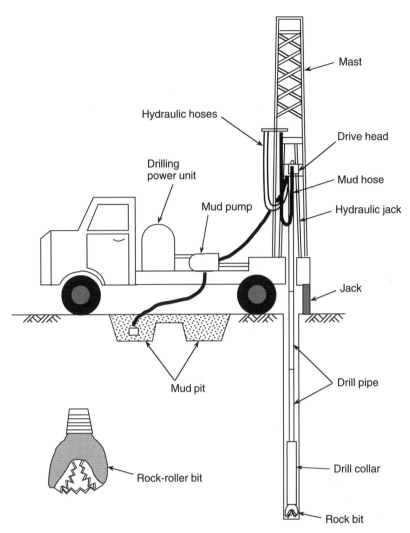

Mast

Hydraulic hoses

Drive head

Drilling power unit

Mud hose

Mud pump

Hydraulic jack

Jack

Mud pit

Drill pipe

Rock-roller bit

Drill collar

Rock bit

Figure 5.16 Rotary drilling is achieved by rotating a drill bit on the end of a series of hollow steel tubes known as drill pipe. Most rigs use hydraulic motors to rotate the drill rods and also to move the drill step up or down the hole. Rigs are usually lorry-mounted and always have a mast to enable the rods to be handled in a vertical plane. A rotary drill bit consists of three sets of wheels which have teeth-like cog wheels (see inset). As the drilling bit rotates, it pulverises the rock into small pieces. The cuttings are removed by a fluid which is pumped down inside the drill stem so that it flows upwards carrying the cuttings to the surface. In water borehole drilling the fluids generally used are air or water. Good progress requires the correct equipment and an up-hole velocity which is high enough to carry the cuttings. Removal of the cuttings can sometimes be helped by using a viscous material like drilling mud. Although muds are commonly used when drilling for oil and gas, they should be avoided if possible for water construction because they tend to block up the aquifer immediately around the borehole. This clogging can significantly reduce the aquifer permeability so that little or no water can flow into the borehole. Drilling muds may also cause water quality problems

There are several different types of rotary drilling; the most common method in use for water well construction being top drive rotary. All rotary methods involve the drill bit being rotated in the borehole to make the cutting action. The bit is attached to a drill string made up of a series of drill rods which are screwed together. These rods are gripped at the top by the rotating part of the drilling rig. In principle, it is exactly the same as your DIY electric drill. A rotary drilling bit is shown in Figure 5.16 together with the other equipment used in rotary drilling. The photograph in Figure 5.17 gives you an idea of what this equipment actually looks like.

Over the last decade or so a drilling foam has been introduced into water well construction and consists of a proprietary foaming agent rather like washing-up liquid which is added to water. Foam is produced by pumping in air from a compressor and the result looks like shaving foam. This frothy mixture is pumped down the hole in the normal way and is very efficient in bringing out the cuttings. It does not affect the performance of the finished borehole and in many ways is better than either air or water. It is best to agree beforehand with your contractor that he is only to use air, water or foam, unless you give him permission in writing to use something else. Do not be tempted to specify exactly which fluid he should use or else if things go wrong he might blame it all on you!

Another form of rotary drilling which you may come across is called down-hole hammer. This method uses a different type of drill bit to normal rotary drilling. A down-hole hammer bit is made to vibrate rapidly when compressed air is pumped through it from the drill pipe. It works very much like the pneumatic jack-hammers that are used to break out roads and concrete. This method has dramatically high drilling rates and may enable a borehole to be completed in just one day. It has the further advantage of coping with the hardest rock. For these reasons it is being used by an increasing number of water well drilling contractors.

Development of Yield

All drilling methods tend to block up some of the pores and fissures in the rock which will supply water to a new borehole. Rotary drilling grinds the rock away, some of it to a very fine powder. The high pressures needed to pump these cuttings up to the surface, force some

Figure 5.17 The photograph shows a rotary drilling rig being used to construct a new well for a farm water supply in Yorkshire, England. At this point in the drilling operation the top of the drill string has almost reached the top of the well casing and the driller will soon add another rod in order to carry on drilling. The drive head which grips and rotates the drill rods can be seen just above the driller's head. The flexible hydraulic hoses which provide the drive power can be clearly seen. The drilling returns which are a mixture of water and rock cuttings are flowing from the well along a short channel to the mud pit where the cuttings settle out. Behind the driller, to the right of the picture, additional drills rods can be seen resting on trestles. (Photograph courtesy of Dales Water Services, Ripon, North Yorkshire)

of them into the rock. Percussive drilling methods also produce a fine rock powder which is smeared on the borehole walls by the up-and-down action of drilling. Unless this material is cleaned away, the borehole will only produce a fraction of its potential yield.

Much of the drilling debris can be removed by pumping, and air-lift techniques are often used for cleaning. Sometimes boreholes are surged. A device rather like a plunger (see Figure 5.18) is moved up and down the borehole, forcing water in and out of the rock as it passes. In this way the fine material is gradually washed out of the rock and finally cleared from the borehole by air-lift pumping.

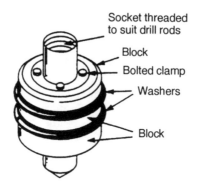

Socket threaded
to suit drill rods

Block

Bolted clamp

Washers

Block

Figure 5.18 A surge block used to clean fine material from a borehole. The block is attached to a drill pipe and moved up and down the borehole. On the downstroke it forces water into the well face and out of it on the upstroke. In some ways it resembles the action of a plunger used to unblock a kitchen sink. This technique can be very effective in pulling fine-grained aquifer material into the well and removing the debris left over from drilling the borehole. (Reproduced from Clark 1988 by permission of John Wiley & Sons)

Boreholes in chalk and occasionally in some other rocks, especially limestone, need to be developed using acid. Most chalk boreholes are drilled by percussion methods which leave a thick smear of chalky mud on the borehole walls. Pumping and surging hardly touch this material but, as it is largely made up of chalk (calcium carbonate), it can be removed by chemical means. This is very similar to cleaning fur from an electric kettle. Hydrochloric acid is injected in the borehole and left to allow the chemical reaction to take place. Sometimes the reaction is helped by gently surging the borehole. The borehole is cleared of spent acid and other debris by air-lift pumping until the water is clean and no longer acid. Most established water well drilling contractors, especially those who work in Chalkland areas, are experienced in these techniques and use them as a standard part of borehole construction.

Another development technique that uses chemicals is employed when drilling in rocks which contain clay, such as sandstone. During drilling the clay becomes smeared on the well face and is difficult to remove with a plunger or by pumping. The chemicals used are polyphosphates such as Calgon, the tradename for a chemical used in water treatment and as a corrosion inhibitor. These chemicals work by increasing the negative charge on the clay particles, which causes them to be forced apart in the same way that magnets repel each other. Dosage rates of $10-40 \text{ kg/m}^3$ are injected through pipes into the borehole. The mixture is left overnight before being pumped out, bringing the clay with it. Care must be taken to dispose of this water without causing an environmental hazard.

Protection from Pollution

Whichever drilling method is used there are certain measures which must be taken to protect your borehole from pollution. These precautions are similar to those taken to protect dug wells. The borehole casing should be sealed into the ground by cement grout being injected into the annular space between the casing and the ground. At the surface the borehole should be surrounded by a concrete pad and covered by a watertight chamber like the one shown in Figure 5.19. Water must be prevented from entering this chamber or it may flow down your borehole, polluting your water supply. The general comments about water wells also apply to boreholes with respect to the proximity of a water source to septic tanks, sewers and the like.

Boreholes which are drilled into sands, and other unconsolidated material which is likely to collapse, need to be supported by casing which extends to the base of the hole. The casing needs to be perforated to allow water to flow into the borehole and this is usually achieved by sawn slots cut by the manufacturer. The size of these slots should be selected so that very little of the aquifer itself can enter the borehole. This is done by the driller and is based on the grain sizes in the aquifer.

Sometimes it is necessary to install a gravel pack round the perforated casing (see Figure 5.20). A gravel pack is really an artificial aquifer and needs careful design and installation. They are most important where the aquifer consists of very fine sand. These sand grains are fine enough to enter the borehole through the smallest slots in the proprietary slotted casing. A gravel pack acts as a filter preventing fine aquifer material from getting into the borehole. The size of the material in the pack is based on the size of the material in the aquifer. Samples of aquifer are taken from the borehole and are sieved to give this information. Your contractor will arrange for the necessary laboratory tests to be carried out so that he can design the gravel pack.

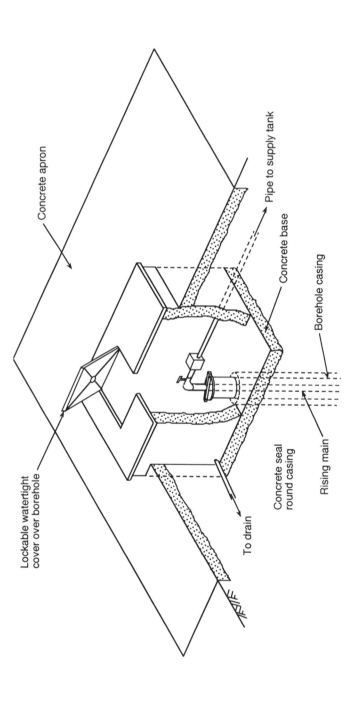

Concrete apron

Lockable watertight
cover over borehole

To drain

Concrete seal
round casing

Rising main

Borehole casing

Concrete base

Pipe to supply tank

Figure 5.19 It is important to protect your well head so that the water supply is not liable to pollution problems or to be accidentally damaged. A typical arrangement is shown in the picture where the top of the borehole is surrounded by a concrete chamber. Alternative arrangements include the use of large-diameter concrete pipes to make the chamber. The well head should be surrounded with a concrete apron which slopes away from the borehole. The top of the chamber should be well above the surrounding area so that surface water cannot easily enter it. Access is provided through a watertight lockable cover and the chamber is drained with a pipe to a ditch which is fitted with a vermin-proof mesh. The drain is important to prevent water from building up and pouring into the well. I have seen flooded chambers on numerous occasions with a water level at the same height as the flange joint on the top of the casing. Sometimes it is difficult to convince the well owner that this level is not a coincidence and that potentially polluting water really is flowing into the well

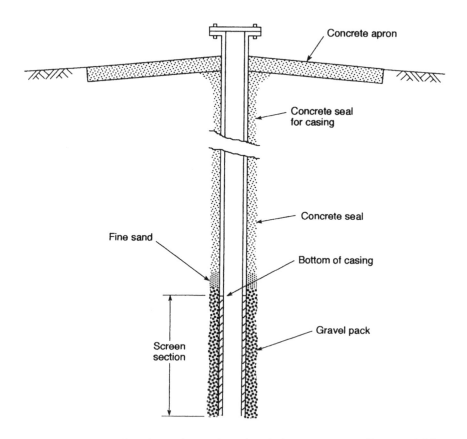

Figure 5.20 A gravel pack may be used in a borehole to prevent aquifer material from entering the well. The grain size of the pack material is carefully selected using information on the grain size of the aquifer obtained by testing samples obtained during the drilling process. A gravel pack should be at least 75 mm in thickness, but there are no advantages for it to be greater than 150 mm. It should be made of clean, well rounded sand and gravel. Siliceous material, that is quartz sands and gravels or flint gravels, are best; processed limestones and limestone chippings should be avoided as they will gradually fuse and allow no water movement. The pack material is placed at the bottom of the borehole on the outside of the screen section using tremie pipes. A layer of sand is placed above the gravel to prevent the grout from invading the pack. The rest of the borehole is sealed with concrete and a concrete pad is built around the borehole top

Borehole Casing

Perhaps one of the most controversial aspects of drilling a borehole is the material to be used to line it because some less reputable drilling contractors may use unsuitable pipes. For example, these days sewer pipe tends to be made out of uPVC, comes in 6 m lengths and is cheap and easily available. Consequently it has been used to line boreholes drilled for private supplies. Unfortunately, it is not made to the same exacting standards as plastic pipe used for water supplies and it is possible that small amounts of toxic substances may leach out of the uPVC into the water supply. Additionally, the wall thickness of the sewer pipe is less than that of purpose-made plastic well casing, resulting in the possible collapse and failure of the borehole. Many countries actually have regulations which cover the acceptable materials that can be used for casing. In the USA, for example, steel casing is fully specified by the American Petroleum Institute (API) Specification, 5A 1984. In the UK there is a British Standard (BS 879) which defines the types of mild steel pipe and plastic pipe that are suitable for lining boreholes. The British Standard specifies three types of joints for steel pipes: welded joints, screw joints and screw socketed joints. The best type by far are the screw joints, but not those with sockets.

During the late 1970s several manufacturers of plastic pipes introduced screw-jointed plastic water well casing into their range. The threads and sizes of these plastic well casings allow them to be used in conjunction with steel casing. There are a number of advantages in using plastic piping. It is cheaper than steel and, being very light compared with steel, these casings are easy to install. The plastics are the same as used in the water industry and are safe to use for drinking water supplies; these plastic pipes are now covered by the same British Standard as steel casing. To all intents and purposes these materials are chemically inert, i.e. plastic casing will not corrode and the life of a borehole can be greatly extended. Most water well contractors have experience of installing this casing and I suggest that you use it.

Make sure that a contractor uses proprietary plastic water well casing and not uPVC sewer pipe or cable ducting which will most likely break and cause many problems. The best plastic casings are made from thermoplastics which are far superior to uPVC. They are more expensive but well worth the extra money.

Well Screens

The manufacturers of plastic water well casing include in their range slotted casing with standard sized slots. In most cases you will be able to use this material without any disadvantage.

There are a number of water well screens designed and manufactured to give optimum hydraulic conditions in a borehole. Some are made of wire wound round vertical rods in one continuous length, rather like a giant spring. Usually they are made of stainless steel, although some firms make them from galvanised steel. This material can have serious disadvantages and generally should not be used. These screens are also available made from plastic.

A further type of water well screen consists of plastic water well casing which has been perforated with round holes, and wrapped around with a special fabric. This material is woven out of plastic thread in such a way that predetermined, very small openings are

Figure 5.21 A selection of plastic pipes used in water well construction. The solid pipes are used as casing and resemble the steel pipes which may also be used. The perforated pipes are used as well screens. The picture also includes two wire-wound screens and a perforated pipe with a gravel pack already stuck to it. (Photograph: Rick Brassington)

produced. The size of the openings is matched to the grain size of the aquifer and this type of screen is reputed to be effective in fine-grained materials.

Other special well screens have louvre-shaped slots, while others have a gravel pack already bonded on. They are all much more expensive than standard slotted casing. There are certain instances where their use in boreholes for small private water supplies is justified but because of their high cost these are few and far between. A variety of well screens are shown in Figure 5.21.

6
Water Quality and Treatment

Before using a source of water for the first time, you will want to be happy that it is safe to drink. This will apply whether it is the supply to a property you have just purchased or a new source you have built. There are three aspects of the water's quality which should be examined. These are the concentration of dissolved minerals, the bacterial content and whether or not it is polluted. Ideally you want a relatively low level of dissolved minerals as this makes the water more palatable, but you certainly do not want any harmful bacteria in the water.

The best and cheapest way to avoid all water quality problems is to select the right type of water source. Build it in a place which is well away from potential sources of pollution and to a high standard, which will stop contaminated water getting into it. Then maintain it regularly! The most trouble-free sources are boreholes, springs and wells. Direct intakes from streams, rivers or ponds are the ones most likely to be at risk of pollution and should be avoided if possible.

The way to satisfy yourself that the water quality is satisfactory is to have it examined in a laboratory. If you have just had a new borehole drilled your contractor will advise you about having an analysis carried out. Otherwise contact the Public Health or Environmental Health Department in your local council offices. It is part of their job to ensure that water supplies do not contain anything that would be harmful to the people drinking it. They will be able to

Table 6.1 Comparison of water quality guidelines and standards for various countries

Parameter	Unit	EEC Directive gl	EEC Directive mac	Canada	Denmark gl	Denmark mac	France	Ireland	UK	USA	WHO
Colour	mg/l Pt–Co	1	20	15	5	15	15	20	20	15	15
Turbidity	Formazin units	0.4	4	5	0.3	0.5	2	4	4	5	5
Odour	Dilution number at 25 °C	–	3	–	–	3	3	3	3	3	–
Taste	Dilution number at 25 °C	–	3	–	–	3	3	3	3	–	NEC
Temperature	°C	12	25	–	–	12	25	25	25	–	–
Hydrogen ion	pH units	6.5–8.5	–	6.5–8.5	7.0–8.0	8.5	6.5–9.0	6.0–9.0	5.5–9.5	6.5–8.5	6.5–8.5
Conductivity	μS/cm at 20 °C	400	–	–	>300	–	–	1500	–	–	–
Chloride	Cl mg/l	25	–	250	50	300	–	250	400	250	250
Sulphate	SO_4 mg/l	25	250	500	50	250	–	250	250	250	400
Silica	SiO_2 mg/l	–	–	–	–	–	–	–	–	–	–
Calcium	Ca mg/l	100	–	–	–	–	–	200	–	–	–
Magnesium	Mg mg/l	30	50	–	30	50	50	50	50	–	–
Sodium	Na mg/l	20	150	–	20	175	120	150	150	–	200
Potassium	K mg/l	10	12	–	–	10	12	12	12	–	–
Aluminium	Al mg/l	0.05	0.2	–	0.05	0.2	0.2	0.2	0.2	–	0.2
Dry residuals	mg/l	–	1500	–	1500	–	–	–	1000	–	–
Nitrates	NO_3 mg/l	25	50	10(as N)	–	50	50	50	50	10(as N)	50
Nitrites	NO_2 mg/l	–	0.1	1.0(as N)	–	0.1	0.1	0.1	0.1	1(as N)	3
Ammonium	NH_4 mg/l	0.05	0.5	–	0.05	0.5	0.5	0.3	0.5	–	–
Kjeldahl nitrogen	N mg/l	–	1	–	–	1	2	1	1	–	–
Permanganate value	O_2 mg/l	2	5	–	1.5	3	5	5	5	–	–
Hydrogen sulphide	S mg/l	–	UO	0.05 (as H_2S)	–	UO	UO	UO	–	–	UO

Substance	Dry residue mg/l	0.1	—	0.1	—	—	NSI	—	—	—
Substances extractable in chloroform	μg/l	0.1								
Hydrocarbons			10		10	10	10	10		
Phenols	C₆H₅OH μg/l		0.5		0.5	0.5	0.5	0.5		
Boron	B μg/l	1000		5000	1000	2000	1000	2000		
Surfactants	μg/l as lauryl sulphate		200		100	200	200	200		
Iron	Fe μg/l	50	200	300	200	200	200	200	300	300
Manganese	Mn μg/l	20	50	50	50	50	50	50	100	100
Copper	Cu μg/l (at outlet)	100			100	100	500	100		
	(after standing)	3000		1000	3000	3000	3000	3000	1000	1000
Zinc	Zn μg/l (at outlet)	100			100	100	1000	100		
	(after standing)	5000		5000	3000	5000	5000	5000	5000	5000
Phosphorus	P₂O₅ μg/l	400	5000		687	5000	2200	5000		
Fluoride	F μg/l		1500	1500	1500	1500	1500	1500	4000	1500
Cobalt	Co μg/l									
Barium	Ba μg/l	100		1000		500	1000	500	2000	
Silver	Ag μg/l		10	50	10	10	10	10	50	50
Arsenic	As μg/l		50	50	50	50	50	50	50	5
Cadmium	Cd μg/l		5	5	5	5	5	5	5	5
Cyanides	CN μg/l		50	200	50	50	50	50	200	100
Chromium	Cr μg/l		50	50	50	50	50	50	120	50
Mercury	Hg μg/l		1	1	1	1	1	1	2	1
Nickel	Ni μg/l		50		50	50	50	50	100	100
Lead	Pb μg/l		50	50	50	50	50	50	20	20
Antimony	Sb μg/l		10		10	10	10	10	6	10

continued

Table 6.1 (continued)

Parameter	Unit	EEC Directive		Canada	Denmark		France	Ireland	UK	USA	WHO
		gl	mac		gl	mac					
Selenium	Se µg/l	—	10	10	—	10	10	10	10	5	10
Pesticides											
per substance	µg/l	—	0.1	—	—	0.1	0.1	0.1	0.1	—	—
Aldrin and dieldrin	µg/l	—		0.7	—	0.03	0.03	—	—	—	0.03
Hexachlorobenzine	µg/l	—		—	—	0.01	0.01	—	—	—	—
Total	µg/l	—	0.5	—	—	0.5	0.5	0.5	0.5	—	—
Total coliforms	number/100 ml	—	0	10	—	—	0.01	0	0	0	0.01
Faecal	number/100 ml	—	0	0	—	—	0	0	0	0	0
Dry residuals coliforms	mg/l	—	1500	—	1500	—	—	—	1000	—	—
Faecal streptococci	number/100 ml	—	—	—	—	—	—	—	—	—	—
Sulphite-reducing clostridia	number/20 ml	—	≤1	—	—	—	0	<1	<1	—	0
Minimum required concentrations											
Total hardness	CaCO$_3$ mg/l	60	60	—	—	—	—	—	60	—	—
Alkalinity	HCO$_3$ mg/l	30	30	—	—	—	—	>100	30	—	—
Dissolved oxygen	O$_2$ mg/l	—	—	—	—	—	—	—	—	—	—

The concentrations or values shown in this table are the maximum permitted values except for the EEC and Danish figures which show guideline (GL) values together with maximum acceptable concentrations (MAC). (Reproduced by permission of John Wiley & Sons).
NEC = No effect on consumers
UO = Undetectable organoleptically
— = No value available

Table 6.2 Water quality guidelines for crop irrigation. (Reproduced from Canadian Water Quality Guidelines, 1987 by permission of the Minister of Supply and Services, 1995)

Bacterial (number per 100 ml)	
Pathogen	Guideline
Animal pathogens	
faecal coliforms	100
total coliforms	1000
Plant pathogens	no guideline

Major ions	
Ion	Guideline
Bicarbonate	no guideline
Chloride	100–700 mg/litre (depending on crop)
Salinity	500–3500 mg/litre (depending on crop)
(Sodium concentration varies significantly with the crop)	

Metals and trace minerals (mg/litre)

Ion	Guidelines	
	All soils	Neutral to acid soils
Aluminium	5.0	20.0
Arsenic	0.1	2.0
Beryllium	0.1	0.5
Boron	0.5–0.6	no value
Cadmium	0.01	no value
Chromium	0.1	no value
Cobalt	0.05	5.0
Copper	0.2–1.0	5.0
Fluoride	1.0	15.0
Iron	5.0	20.0
Lead	0.2	2.0
Lithium	2.5	no value
Manganese	0.2	10.0
Mercury	no value	no value
Molybdenum	0.1–0.05	0.05
Nickel	0.2	2.0
Selenium	0.2–0.05	no value
Uranium	0.01	0.1
Vanadium	0.1	1.0
Zinc	1.0–5.0	no value

advise you on suitable laboratories even if they cannot carry out the work for you. Another source of information on appropriate laboratories will be water supply companies or authorities. It is important to use a laboratory that is experienced in water analysis, so check this out before having any work done by asking them to list some of their water clients.

The water analysis should include a wide range of substances and provide information to satisfy the regulations which govern the quality of water for human consumption or use in agriculture or manufacturing. These regulations are discussed at the end of this chapter and information on the water quality standards is provided in Tables 6.1 and 6.2. Do not be tempted to think that you will only need to have one analysis carried out on your water supply. It is good practice to have checks carried out every year or so and in some cases to look for water quality changes which may occur on a seasonal basis.

WATER QUALITY CHARACTERISTICS

When water samples are examined for their bacterial content, the main tests are to look for the various organisms which live in sewage and animal excrement. Not all these bugs cause disease but the presence of one type often means that others are there, some of which can cause very serious illness. Happily, if your source has been properly constructed and maintained, it is unlikely to have any harmful bacteria in it. If in doubt, boiling the water for a quarter of an hour will kill any germs. Unfortunately, boiling will concentrate dissolved minerals and should not be used if you have nitrate or similar problems.

The chemical examination of water reveals the dissolved minerals in the water. Most people are familiar with water being *hard* or *soft*. Hard water has a relatively high level of minerals dissolved in it, while on the other hand soft water is relatively free from such dissolved minerals. You can easily recognise soft and hard water by the ease with which you can lather soap. Hard water does not readily form a lather and often leaves a scum in the basin.

Another feature of hard water is the furring of kettles and pipework. This is caused by *temporary hardness* which is produced by the presence of calcium and/or magnesium carbonates. These minerals precipitate out of solution fairly readily and produce the deposits in kettles and pipes. *Permanent hardness,* on the other hand, is caused by the presence of calcium and magnesium sulphates which stay in solution. One of the advantages of hard water is that the coating inside pipes protects them from acid water attack. Also hard water tends to be more palatable than soft water which may taste "flat".

The sulphate content of water can also be in the form of potassium or sodium sulphates as well as the calcium and magnesium sulphates which cause permanent hardness. High concentrations of potassium and sodium sulphates may result from the rainfall being affected by industrial atmospheric pollution which produces high levels of sulphur dioxide in the rain. Where it occurs in groundwater, it may also originate from minerals being dissolved from the rocks. Large amounts of these sulphates may give a bitter taste to the water and cause diarrhoea. Sulphuretted hydrogen (hydrogen sulphide) is a gas which gives an unpleasant smell to some well waters. The smell usually disappears quickly when the water is exposed to the atmosphere and most such waters are otherwise quite pure.

Carbonate and sulphate minerals are not the only chemicals frequently found in water. There are often low concentrations of sodium chloride (common salt) and other chlorides. There can be nitrates, usually originating from agricultural fertilisers or sewage. Sometimes there are various metals dissolved naturally in the water, but these are usually in very low concentrations. The one which most commonly causes problems is iron, closely followed by manganese, but other metals such as lead, copper and zinc can be dissolved from pipes and tanks by acid waters and then present difficulties. If there are concentrations of iron in water of 0.5 mg/l or more, the iron may steadily come out of solution as the water stands in tanks and pipes. This will produce unpleasant black particles in the water and will cause rust stains on clothes, sinks and toilets, beside giving a bitter taste. Manganese causes similar problems, including precipitation of black particles which may cause staining, and also gives an unacceptable taste.

Another important property of water that is to be used for drinking is its pH, which is a measure of the concentration of the hydrogen ion. It is used as a measure of the acidity of substances. Water with a pH of 7 is said to be neutral; if it is less than 7 the water is acid; and if it is more than 7 the water is alkaline. The World Health Organisation recommends that the limits for the pH of drinking water are in the range 7.0 to 8.5. If the value falls outside this range but lies between 6.5 and 9.2, then it is still acceptable although I would give thought to treatment. Any other pH values are likely to give rise to serious problems.

The pH is measured on a scale of 1 to 14 but it is important to remember that this is not an ordinary scale, like that used to measure temperature for example. The pH scale is logarithmic which means that each step is tenfold greater than the last one. For example pH 7 is ten times more alkaline than pH 6. Similarly, pH 8 is 1000 times more alkaline than pH 5. Conversely, pH 6 is 100 times more acid than pH 8, and so on.

Portable hand-held instruments are available to measure the pH of water in the field. They are straightforward to use but care is needed to calibrate the instrument using a standard solution of known values. They are costly to buy (around US$ 400–500) but can be hired from instrument supply companies. A relatively cheap alternative, which will be adequate in most instances, is to use pH indicator paper.

Most people will remember using litmus paper to test whether a solution is acid or alkaline in school chemistry lessons. Deep colours indicate the degree of the acidity or alkalinity. The pH value can be assessed by using an indicator paper which is similar to litmus paper but turns a different colour with each pH range. The colour of the paper is compared against a chart supplied with it so that the pH can be easily determined. Indicator papers can be obtained which cover a wide range such as pH 1 to pH 10, or a more limited range of say pH 6 to pH 8. Those papers with a limited range still use the same ten colours so enable a more accurate estimate of the pH to be made. It should be possible to buy pH indicator paper from scientific equipment suppliers but if you get stuck the science teacher at your local high school is likely to be able to give you an address.

It is important not to have a very acid water because these waters may dissolve metal from pipes and tanks which may cause a couple of problems. First, you will be drinking these metals and they can cause serious illness, and secondly, your tanks and pipework will soon become full of holes and need major repairs.

A large number of cases have been recorded which have been caused by low pH water dissolving lead from pipes and lead-lined tanks. Health problems caused by high concentrations of lead in drinking water have received a lot of publicity over the years as high concentrations of lead can cause brain damage, especially in children. The best method of coping with this problem is to treat your water so that it has a pH between 7 and 8.5 or to take all the lead pipes out of your house. You should treat the pH anyway if it is more acid than about 6.5. Treatment of very acid waters is important as in extreme cases they can increase tooth decay and may even cause stomach ulcers.

An acid water supply may dissolve copper from your pipes and hot water cylinder and zinc from galvanised iron tanks. The answer to this problem is the same as for high lead levels. Treat your water to produce a neutral pH or use non-metallic pipes and tanks for non-drinking supplies where the acidity does not matter. Some manufacturers even make PVC pipes for hot water systems but you still need to use copper pipes near boilers and a copper hot water cylinder.

The scale deposited inside pipes by hard water will prevent them from being dissolved. Unfortunately hard water and high pH levels tend to go together so if you have aggressive water the chances are it will be soft and no protective layer will form inside your pipes.

Coloured water can be a frequent problem with private water supplies in areas where there is peat. The most usual colours are yellow and brown and indicate the presence of organic material in the water. Waters which have a high iron content may also be coloured. In many cases the colour is apparent rather than actual and is caused by minute particles suspended in the water. If filters are used or the water is treated with a chlorine solution, the colour is often removed.

The use of pesticides in agriculture over the past few decades has resulted in small concentrations of these chemicals being present in a high proportion of waters. The analysis for these materials is very specialised and tends to be expensive. If in any doubt, however, it is worth having your water supply checked for the types of chemicals which are used by the farmers in your area. Farmers' organisations or government agricultural departments are likely to have records of the types of chemicals used in your district. This will help decide on the list of chemicals to look for in the analytical tests. Treatment to remove these substances is often difficult and may involve activated carbon or reverse osmosis (see below).

Most people will have noticed that the clear water in a stream quickly becomes coloured as the flows increase in response to heavy rain which washes soil and other materials into the stream. Other similar processes can change water quality on a short-term or seasonal basis. Tests carried out recently on several springs in part of northern England show that they are contaminated by bacteria originating from sheep droppings being washed into the sources by rain. There is a time-lag of up to three days from the rain falling to the otherwise bacteria-free water being polluted. A similar pattern was noticed with iron and manganese concentrations which increase after rain as the percolating water dissolves these minerals out of the soil. Other changes were also noticed although these seem to have a different cause. The concentration of aluminium varied in accordance with the degree of acidity of the rain because those rain showers which were very acidic dissolved more aluminium from the soil. Lead concentrations also varied with time but these were related to changes in atmospheric pollution largely originating from vehicle exhaust gases. The message from this work is that you should expect the quality of your water to vary and a number of samples are needed to ensure that the water quality is properly understood to enable appropriate treatment methods to be selected. These comments apply to surface water sources, springs and shallow wells. It is unlikely that deep-seated groundwater would be effected in this way.

There are a number of micro-organisms which can effect human health and can be transmitted through drinking water. They can be divided into three main groups: *bacteria*, *viruses* and *protozoa*. Bacteria are the most important of these little "bugs" in terms of the frequency with which they are found in drinking water and identified as the cause of outbreaks of disease. Most of these are connected with faecal contamination and give rise to dysentery, typhoid, paratyphoid, cholera and tuberculosis. *Legionella* is a well-known bacterium which causes a severe form of pneumonia. It is clearly worthwhile ensuring that your water supply is free from these creatures, so build it correctly to stop them getting in.

Viruses are pretty tough things. Most of those which affect humans are excreted when visiting the lavatory. The viruses can survive in the environment for very long periods but they can only reproduce when they have managed to get into a host cell. As a result there are lots of them likely to find their way into your water supply. As with bacteria, the best thing to do is to protect your source so that the chances of contamination are small.

Protozoa are simple single-celled animals that have been around since the early days of life on earth. This means that these are tough cookies, too. The main ones which are commonly found in drinking water and may cause disease are *Cryptosporidium* and *Giardia*. Both *Cryptosporidium* and *Giardia* are parasites which are widely found in nature and infect a very wide range of animals including pets and farm animals as well as human beings. At times in its life-cycle a protozoan will form a sort of protective shell and can survive in water for a long time while waiting for a host animal or person to come along. They are also able to complete their whole life-cycle within a single host which means that once infected you stay that way for life. Contamination is again via faeces, human or animal, so keep it out of your water!

TREATMENT

Water treatment is expensive in both time and money and it is always worth considering an alternative water supply before installing large-scale treatment equipment. Many forms of treatment use chemicals and these must be kept topped up, otherwise you will be drinking problem water again. Some treatment equipment needs regular routine maintenance, so one way or another you can expect to give water treatment some attention at least every few days.

Sometimes, however, it is not possible to avoid some form of water treatment. For example, in some parts of the Pennine Hills in northern England the most reliable supplies are from boreholes but the groundwater tends to have a high temporary (carbonate) hardness and may also have high iron concentrations of up to 8 or 10 mg/l. This water needs treatment to soften it and remove the iron.

In this section the most usual water treatment processes are explained but if your water has an unusual quality problem you should get expert advice on the correct treatment method to use. The companies which make treatment equipment and chemicals usually provide this service for their customers.

Bacterial Treatment

Private water supplies often contain bacteria which do not harm the people who drink it regularly. The world famous mountaineer,

Chris Bonnington told how on one expedition to the Himalayas he was the only member of the party, apart from the local guides, who did not suffer with diarrhoea caused by their water supply. He attributed it to having a spring supply to his home in the Lake District, England. Everyone has a population of bacteria which live inside their intestines. They do no harm, and in fact are essential in digesting our food. If we drink water containing some bacteria it may alter the natural balance and cause diarrhoea and other stomach upsets. By drinking untreated water, Bonnington had built up a natural resistance to water-borne bacteria and so he came to no harm. His mains-water drinking colleagues, on the other hand, were used to drinking disinfected water and suffered for it. This does not mean, however, that you should not include some form of precautionary treatment in your water supply system and I recommend that in almost every instance, one or other of the systems described here should be used.

Water which is contaminated by faecal bacteria should not be used for domestic purposes. The treatment methods described here are precautions where a supply is liable to periodic bacterial pollution and are not ways of treating badly polluted water.

The water supply industry deals with this problem by using chlorine gas to kill off any bacteria which may be in the water. Sometimes the dose used is a bit too high and a chlorine smell remains with the water right up to the tap. Most people will have smelled this chlorine in their mains water supply at some time or another. The chlorine is bled slowly into the water from storage cylinders. This is a hazardous operation and is not suitable for private water supplies.

Chlorination using chemicals which contain chlorine, rather than the gas, is one way of disinfecting larger water supplies but is both expensive and tedious if only small quantities are involved. The water supply for a larger establishment like a school or hotel may be treated using a ready-made chlorine solution. This is fed into the water supply using relatively simple apparatus. Before deciding on chlorine-based treatment, however, think about using ultraviolet light as discussed below.

The most easily available chlorine solutions are those made with sodium hypochlorite. There are a large number of them on the market, sold as household disinfectants under various trade names such as Chloros, Domestos, Voxsan and Parazone. If in doubt read the label on the bottle; it should clearly state the chemicals it contains. They have chlorine concentrations ranging between 5 and 10% by weight although they may lose about half their strength if stored for about a year. The losses are much greater if the solution is exposed to the air or sunlight.

Another source of chlorine is calcium hypochlorite. This is a white granular material that is usually sold in tablet form for ease of handling. There are a number of brand names such as Pit-Tabs, HTH Tablets and Chlor-Tabs. Over two-thirds of its weight is available chlorine and it can be stored in sealed containers as a powder for many years without losing its strength. It can be made into solution by dissolving in water although great care must be taken to protect skin, eyes and clothes. First mix the weighed powder with a little water to form a cream, then add further water to produce a solution of the desired strength.

The chlorine solutions are drip-fed into the water supply at a rate which depends on the strength of the solution and the water use. Detailed instructions are supplied by the manufacturer of the equipment so that it can be used efficiently.

Disinfection

Bleaching powder (chloride of lime) can be used for disinfection but should not be used for routine chlorination of water supplies. Bleach is a mixture of calcium hypochlorite and calcium chloride and contains about one-third free chlorine by weight. It can be used to disinfect new sources such as spring collection chambers and wells, besides cleaning new pipes and tanks in your distribution system. It is very important that you only use chlorine-based disinfectants. Others are phenol based which will cause long-term taste problems and must not be used under any circumstances.

For disinfection use a solution which contains about 20 mg/l of chlorine. This means diluting 4 ml of 5% sodium hypochlorite solution in 10 litres of water. When using bleaching powder dissolve about 80 g (about a cupful) in 10 litres of water. Allow the lime sludge to settle out and the liquid will be a 0.2% solution. Dilute this by adding 100 ml to 10 litres of water to produce a 20 mg/l strength solution. Take great care when handling chlorine-based powders and solutions. They are very caustic and can burn your skin and damage your eyes. Wear protective gloves and an overall, and protect your eyes with goggles or glasses.

When disinfecting a new part of your water supply system, fill the catchpit, well, tank or pipeline with a chlorine solution of 100–200 mg/l strength to allow for dilution by the water already in the system and make sure that it is thoroughly mixed. Work out how much water is already there and add a solution strong enough to end up with about 20 mg/l once it has been diluted by the water already there.

Leave it in contact for up to 24 hours if possible, or at least overnight, before draining it off. When you disinfect pipework open all the taps until the water has a distinct chlorine smell. The running water may mean that fresh water has entered the tank or chamber feeding the pipes so top it up with extra chlorine solution to maintain the strength. After disinfection make sure that the system is emptied of all water which smells of chlorine. This is likely to entail allowing your spring catchpit to drain several times or pumping your well for several hours.

If you have been disinfecting a new tank or pipeline, drain off the chlorine solution and refill it with clean water. Once this water has been drained off you should have removed all the chlorine solution and can put the system into operation.

Ultraviolet Radiation

Ultraviolet light (UV) has been shown to have a powerful germicidal effect and can be used to kill off almost all micro-organisms in water without having to add chemicals. The UV light is produced from mercury-vapour lamps which produce the same sort of violet glow you see on sunbeds. A UV treatment system consists of one or more lights, depending on the treatment capacity, and has automatic control and safety systems. The instruments used to treat a domestic water supply generally look like a black plastic box about 30 cm long. Figure 6.1 shows the typical arrangement inside the box. They are simple to use: you only need to plug them into the electricity supply and periodically change the bulb at the frequency specified by the manufacturer. UV radiation can cause severe damage to your eyes. This is why you have to wear special goggles when you are using a sunbed to top up your tan. Under no

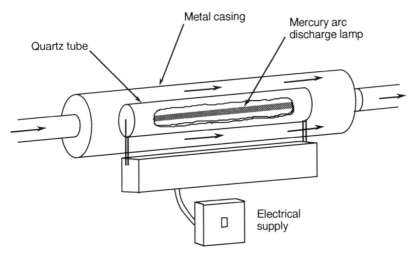

Figure 6.1 A convenient way of disinfecting your water supply is to install an in-line ultraviolet (UV) unit. Disinfection depends on the light penetrating the water and is not effective on very coloured waters. As such it is best suited to groundwater supplies. The UV light is produced from a low- or medium-pressure mercury arc discharge lamp which is housed in a quartz tube. This is in turn placed in a metal tube with a space left for water to flow between the two tubes. The unit is sized according to the flow through it and is designed to give about 1 s exposure to the water as it passes by. The maintenance is easy as it only requires the lamp to be changed on a regular basis. Beware of opening the cover with the unit turned on as the UV light can seriously damage your eyes

circumstances should you ever turn on the light without the protective cover fixed in place.

Filters

Filtration through a cartridge filter is a more satisfactory way of coping with bacterial problems than chlorination when only small quantities of water are involved. A filter will also remove silt and other suspended material from the water and often helps improve water with a poor colour. All private supply systems should have one of these filters on the end of the pipe leading from the water source.

In some ways pre-coated filters resemble an oil filter fitted on a car engine, as you can see from Figure 6.2. There are several designs for these filters. Some consist of a coating material supported on a wire frame and are screwed on to the end of the intake pipe. There

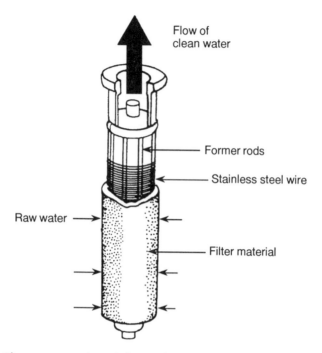

Flow of
clean water

Former rods

Stainless steel wire

Raw water

Filter material

Figure 6.2   There are a number of designs for pre-coated water filters. In the example shown the filter comprises a stainless steel frame which is coated with unglazed ceramic clay. It has small holes which provide the filtering action and is coated with silver which acts as a disinfecting effect. This type of filter is immersed in the untreated water which flows through it as it is pumped into the supply

are three types of coating material used in the manufacture of these filters. Diatomaceous earth is a sort of rock formed out of the remains of microscopic plants called diatoms. It is fine grained and is highly absorbent, so makes excellent filters. Perlite is a glassy volcanic rock which is ground up and used to coat filters. The best material used as a coating for these filters, however, is cellulose.

Porous pot filters look a bit like pre-coated ones but have no supporting rods. They are made from unglazed ceramic material to form a hollow rod. This is open at one end and generally referred to as a "candle". This type of filter is the most convenient sort for small supplies. They can be obtained already impregnated with ionic silver which has a strong sterilising action and will kill any bacteria present in the water. There is some evidence that a silver-impregnated candle can also remove some viruses. The only maintenance they need is occasional washing to remove dirt and other impurities which have collected on the exterior surface of the candle. Many people recommend fitting a silver-impregnated ceramic filter candle as a precautionary measure against pollution.

An alternative design is for the filter action to be made by a single thread of yarn which has been wound round the metal frame. Cotton, wool, rayon, fibreglass, polypropylene (propathene), nylon and other polymers are all used. The simplest ones use a paper filter rather like the air filter on a car engine. The size of the particles that can be removed varies from one filter medium to another and your supplier will advise you on the capabilities of each one. When they are in use these filters slowly clog up and need replacing. As a general rule the larger the pore size, the longer they last. If you need to remove very fine particles it is usual to have a coarse grade filter to pre-treat the water in order to prolong the life of the filters.

Activated carbon is used in water treatment to remove organic pollutants which stick to the carbon on a microscopic level. Cartridge filters filled with activated carbon are also available and an example is given in Figure 6.3.

Water Softening

There are a number of commercially available domestic size softeners. These are sold widely in areas where the main water supply is hard but there is no reason why they should not be used in a private water supply system.

The water treatment process used by these softeners is called "base-exchange". The softener consists of a cylinder partially filled with a carefully manufactured granular resin material. Water is piped into the top of the cylinder under pressure and passes through the resin before it flows out near the base of the softener into the supply system. As water flows through the resin a chemical process replaces calcium and magnesium in the water with sodium from the resin. The process does not reduce the total concentration of dissolved minerals in the water but sodium minerals do not affect lathering with soap so the water appears soft.

This method of softening is generally not used for a drinking water supply. The process increases the sodium levels in the water and although regulations may allow up to 400 mg/l, I suggest that you should make sure that total levels do not exceed 100 mg/l if the

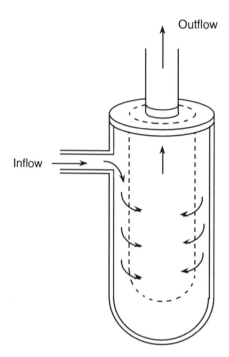

Outflow

Inflow

Figure 6.3   Cartridge filters are fitted into the supply pipe and the water flow is directed through them. They are commonly filled with activated carbon or ion-exchange resins. As the water flows through the filter medium the impurities are either removed by filtration, adsorption or chemical reaction. This type of filter is available to remove dissolved metals, chlorine and organic chemicals. It is important to change them as frequently as the manufacturer recommends as the treatment capacity is limited. Do not try to economise by using a filter longer than recommended; not only do you lose the treatment effect but you may also develop new water quality problems caused by the decay of the organic material which has been caught in the filter

softened water is to be drunk. Quite often the supply to the kitchen tap is not softened so that hard drinking water is still available. This avoids any difficulty caused by increased sodium levels after softening. Water is piped to the kitchen tap from the supply main before the rest of the water flows into the softener. These plumbing arrangements are shown in Figure 6.4.

This method of water softening has a great advantage in that the resin can be regenerated and used over and over again. Regeneration is carried out every few days, depending on the size of the softener and the quantity of water treated. This is usually achieved by passing a solution of common salt (sodium chloride)

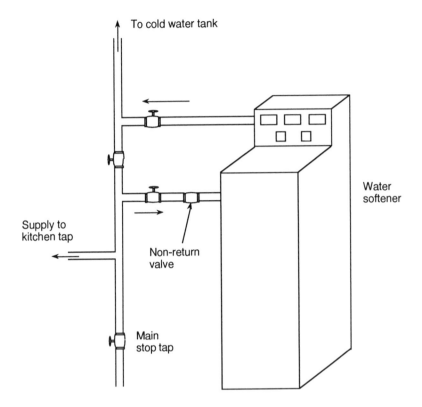

To cold water tank

Supply to
kitchen tap

Non-return
valve

Main
stop tap

Water
softener

Figure 6.4 Most water softeners work by ion exchange involving common salt. Before installing new equipment, check the manufacturer's instructions. Some recommend that you install a pressure-reduction valve in the feed pipe. It is a good idea to include a non-return valve in the feed pipe so that brine cannot be syphoned back into the main pipes. Only soften that part of your water supply which needs it. It is usual not to soften the main drinking supply at the kitchen sink because there is medical evidence that the incidence of heart disease is less amongst hard water drinkers. It is also pointless in softening the supply to your garden or any out-buildings and in flushing your toilet with softened water. To install the water softener, turn off the inlet pipe at the stop-cock. Cut a section out of the main feed pipe and replace it with two tee-pieces separated by a bypass valve as shown. This creates a loop to connect the water softener. Make sure that you can isolate it by installing valves on both the inlet and outlet pipes. The water softener will require an electricity supply but make sure that you isolate it with adequate fuses or trip switches

through the resin, usually in the opposite direction to normal water flow. This exchanges sodium for calcium and magnesium which form insoluble salts which are washed out of the softener as a sludge. Follow the manufacturer's advice on the disposal of this sludge.

pH Adjustment

To ensure that the pH of your water supply is in the range pH 7.0 to 8.5, you can treat it in a similar way to the softening process. The same type of equipment is used but this time granules made of treated dolomite are used instead of the resin. Obviously, this time you would treat all the water before any of it is used and not leave out the drinking water. Dolomite is a type of limestone which is made of calcium and magnesium carbonate. The granules are made so that they slowly dissolve as water passes through them. The process is designed so that treated water has a pH between 7.0 and 7.5. Hardness is increased but not by an amount that is likely to cause any serious worries.

Iron and Manganese Removal

Both iron and manganese may be present in groundwater abstracted from a number of different aquifers including some sands and sandstones, slates, some limestones and in areas where there are peat bogs. The water is generally clear when it is first abstracted but on contact with the air the dissolved iron reacts with the oxygen to form an insoluble form. This then precipitates in the water as a rust-coloured silt. Water which contains significant concentrations of iron usually has an acid pH. Increasing the pH by the process described above will cause the iron to precipitate out of solution. It will produce an insoluble sludge which remains in the treatment cylinder. To make sure that small particles of iron are not carried into the supply system, fit a ceramic filter candle to the draw-off pipe inside the cylinder. Do not forget the trick of a coarse grade filter to pre-treat the water if necessary and remember to clean or change the filter at regular intervals.

Manganese can be removed in the same way although removal is more difficult as it needs a pH of between 8.5 and 9. It is usual to use two stages of filtration, the first removes any turbidity and any residual coagulant chemicals, and the second treats the manganese at the high levels of pH.

Iron can also be encouraged to precipitate if it reacts with dissolved oxygen. Some commercially available iron removal equipment for both small and large supplies works on this principle.

The cylinder is filled with granular filter material which acts as a catalyst; the filter can be cleaned by back washing and will last a long time because it is not used during the iron-removing process. High levels of manganese may be associated with the high iron concentrations. Manganese can be removed by oxidation in just the same way as for iron.

In a few instances of dissolved iron problems there may be complications caused by iron bacteria. There is a group of bacteria that cause the oxidation of iron and manganese, resulting in the formation of glutinous slimes which block pipes, filters and other parts of the water system. The slime may decay causing problems of quality, taste and smell. If you have this problem clean out the slime and disinfect using a strong chlorine solution (20 mg/l). Treatment to remove the iron or manganese will help prevent a further outbreak.

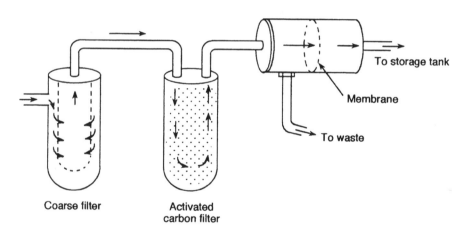

Coarse filter   Activated carbon filter

Figure 6.5 Osmosis is a natural process where water will flow through a semi-permeable membrane which separates two solutions of unequal strength. The flow is in the direction of the stronger solution so that eventually the two are the same. Only water passes through the membrane as the pore sizes are too small for larger molecules. Water treatment by reverse osmosis (RO) has a flow direction opposite to the natural one which is achieved using a pump to pressurise the system. RO filters use a microporous plastic sheet which allows water molecules to pass but retains practically everything else. A typical unit will pre-treat the water with a coarse filter to remove undissolved particles and an activated carbon filter to remove some of the contaminants thereby prolonging the life of the membrane filter. The water flows through the membrane, leaving a concentrated solution behind which requires regular removal. RO systems will treat most waters, but they are slow to operate and very expensive to buy

Reverse Osmosis

Reverse osmosis is a method of producing high quality water from very poor water indeed. Some island water supplies rely on sea-water treated in this way to remove all the dissolved minerals. The process uses a semi-permeable membrane which is effectively a plastic sheet. The semi-permeable membrane has very small holes in it which stop almost all minerals and substances other than water molecules from passing through it. These membrane filters are expensive to run however, and require a lot of "raw" water to produce relatively small quantities of treated water. A domestic-scale reverse osmosis filter is shown in Figure 6.5.

LEGAL CONTROLS ON WATER QUALITY

Drinking water quality is covered by a variety of regulations and standards which specify the maximum concentrations of a long list of substances which could be in a water supply. Perhaps the most widely used guidelines are those published by the World Health Organisation (WHO) as a series of Quality Directives for Drinking Water (WHO, Geneva, 1986). The values quoted by the WHO are recommendations only and it is left to individual countries to set their own standards.

Member states of the European Union are required to meet the standards set down in the European Council Directive of 15th July 1980 relating to the quality of water intended for human consumption (No. 80/778/EEC). This Directive covers all drinking water and that used in the production of food, but not bottled mineral water which is covered by a separate Directive (89/397/EEC). In the drinking water Directive there are Guide Levels (GL) set for 62 parameters which are levels seen as being satisfactory, but there may be circumstances when it is permissible to drink water with higher concentrations. Some of the parameters also have Maximum Admissible Concentrations (MAC) laid down and no waters should be used which exceed these levels. In this way, the Directive specifies the maximum permitted concentrations for all dissolved minerals in the water. It also stipulates the limit of bacteria in water. The general appearance of the water, that is its colour, smell and taste, are all covered. The Directive also spells out how often tests should be carried out to look for all these chemicals and bacteria.

In England and Wales the EU directive is met by the standards proscribed by the Private Water Supply Regulations 1991 and in Scotland by the Private Water Supply (Scotland) Regulations. The responsibility for enforcing these regulations lies with the Environmental Health Departments in the local authorities.

In the USA, the US Environmental Protection Agency has established standards for drinking water. There are many similarities with the WHO and EC standards and the US drinking water standards. All three standards have been included to compile Table 6.1, together with information from a selected number of other countries.

If you want more detailed advice about how water quality regulations affect your water supply contact your local Environmental Health Officer or Medical Officer of Health in the local council office.

QUALITY CRITERIA FOR AGRICULTURE

Generally speaking, if water is suitable for human consumption it is perfectly all right for animals and for use in the dairy. If your water supply has a high concentration of minerals, show your veterinary surgeon or farming adviser the chemical analysis and ask for their help.

Fish farming requires water of high quality. It should have saturation concentrations of dissolved oxygen, otherwise the fish cannot breathe. Spring water or water from boreholes may have very low levels of dissolved oxygen. It can be increased by running the water over a series of cascades which mix air and water, allowing oxygen to be dissolved from the atmosphere. Plants in ponds will increase oxygen levels during the day, but at night respiration of the plants can bring the oxygen content down to dangerously low levels. It is important to have the water pH in the range 7 to 7.5. In view of the large quantities of water needed for a fish farm it would be completely uneconomic to treat the water. A suitable source of water, therefore, will determine the siting of a fish farm.

If you are planning to use water for irrigation the quality is very important. Some salts, particularly when in high concentrations, may harm plant growth by reducing the amount of water they can take up. Others can cause harm by their toxic effect. It is important that you have the water you intend to use for irrigation analysed and if in any doubt, obtain advice on its suitability before you invest a lot of money in the irrigation system. Table 6.2 gives some guidelines on irrigation water quality.

A serious potential problem associated with irrigation is a build-up of salts in the soil, a process usually called *salination*. It is caused by the evaporation of the irrigation water leaving behind the dissolved salts. Besides impacting on the soil structure, the process causes a build-up of the concentration of toxic elements in the water. The most effective way to reduce the problem is to use sprinklers which minimize water use. A variety of other techniques are being tried in many parts of the world where this is a problem. However, this is beyond the scope of this book and you must rely on locally obtained advice.

7

The Rest of the System

The first step in building a water supply system is to decide where you can get the water. Having solved that problem, you should sit down and think about the other parts of the system. This chapter has been included to help you design a water supply system but it is not a guide to plumbing. You should consult a good DIY manual to help you sort out the plumbing system inside your house.

You need to be able to get the water from the source to the tap. The supply system should include some storage to fall back on if there are problems with the water source. Storage is also needed to iron out the major fluctuations in demand. This avoids the pump cutting in every time a tap is turned on and allows several people to have a bath without waiting for the extra water to be pumped up.

To do all these things, a water supply system will need to be made up of pipes and valves. It will have a storage tank or reservoir and, when necessary, a pump. The cheapest water supplies are obtained where pumping can be avoided and water is moved through the pipes by gravity. This is often possible where water is being taken from springs or streams and rivers, but depends on the lie of the land. Where wells or boreholes are used a pump will be needed, except where a borehole is artesian.

If you have a borehole which flows under artesian pressure, fit a valve to prevent if from flowing to waste when you do not need the

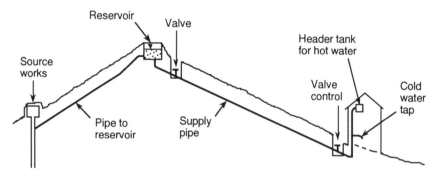

Figure 7.1 This simple diagram demonstrates the essential features required in a water supply system. Water is abstracted from the source and pumped to a reservoir where it has a sufficient pressure head to flow to all parts of the distribution network. Besides giving pressure head, the reservoir also provides storage so that there is an adequate flow to meet fluctuations in demand. The flow of water round the system is controlled using a number of valves which allow sections to be isolated for maintenance. A common alternative to using a reservoir and gravity flow is for water from a well to be pumped into a pressure tank where there is sufficient head to enable water to flow to all parts of the system

water. A continuously flowing borehole causes a gradual reduction of artesian pressure and hence the flow. This means that you will eventually have to use a pump. This waste of water can cause you other problems if the construction of your borehole allows small silt or clay particles to enter it. The reduced flow will allow more and more silt to settle, eventually clogging your borehole and so causing further reductions of artesian flow.

These days pipes and tanks are available which have been made from lightweight non-corrodible materials such as plastics or reinforced glass fibre. The lightness of these materials makes them much easier to install than pipes and tanks made of more traditional iron and steel. They will also last longer as they do not corrode and the system will require less maintenance.

The various elements which are involved in the water supply to a medium-sized farm are shown in Figure 7.1. Obviously there are many permutations of possible arrangements involving different types of source, etc., but this example, based on the water supply to a farm in Devon, England, gives an idea of the things to include in your planning process.

LEVELLING

When you are designing a water supply scheme it is often necessary to measure the relative ground levels of the source, the storage tank or reservoir, and the various outlet points where the water will be used. If you are planning to use a ram pump, for example, you will need to know both the drive head and the delivery head. It is necessary to site a storage reservoir high enough to deliver water under gravity flow to all parts of your farm, or if you are trying to size a pump to deliver water to a tank you will need to know how high the water has to be moved.

The best way is to use a surveyor's level, often called a "dumpy" level, but if you cannot get hold of one, there are several alternative techniques which use easily obtainable equipment. Whichever technique you adopt, including the use of a dumpy level, there are a few standard rules which should be followed. Surveying is always a job for two people and you are likely to find it difficult and get it wrong if you try it on your own. Use a second person to help hold the equipment but make sure that you explain carefully what you expect of them before you start. Frayed tempers do not make for accurate readings! It is important to record all the readings systematically in a notebook or on a note-pad attached to a clipboard. First carefully take the measurement which is involved (usually the height above the ground), write it down, and then re-take the reading to check that you got it right. Do it again if there is a discrepancy. When you come to make your calculations at the end of the levelling exercise, adopt the same principle and double check your arithmetic; after all, it would be silly to get it wrong through a careless error.

Abney Level

An Abney level is the name commonly used for a sighting meter, or inclinometer. It is an instrument which measures the angle of slope between two points. With care these instruments can give acceptable accuracy and are much quicker to use than the other methods I cover here. Figure 7.2 shows how a sighting meter is used and how to calculate heights from the angles you have measured. You will need

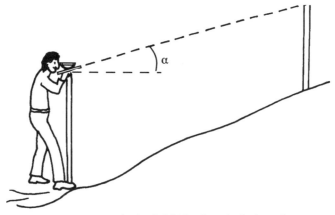

α = Angle of sight line from the horizontal

Figure 7.2 An Abney hand-level is used to measure the angle between two points on the ground so that you can calculate the height difference between them. To use this instrument you must first set up two posts about 1.5 m high, as far apart along the line of your section as it is convenient to measure and so that they are easily seen from each other with the lower one being at the start of your measuring section. Measure the height of the top of each post above the ground and write it in your notebook. If you can, try and make them the same as it will make your calculations easier. Position your hand-level on top of the lower post and sight to the top of the second one and record the angle between them. Write the value in your notebook, then take it again as a check. Move the lower post to above the other one and repeat this sequence of measurements, continuing to the top. The height difference between each point is the distance between the points multiplied by the sine of the angle you measured. Use a calculator with a *sin* function or a set of tables. It is easy to calculate the total height if you take it in stages

a calculator with a sine function (usually abbreviated to "sin") or a set of mathematical tables giving sine values like the ones you may remember using at school.

Spirit Level and Plank

A standard builder's spirit level is only about 900 mm or 1000 mm long but it can be extended to 3–4 m by using it in combination with a plank. It is essential to ensure that the plank is absolutely straight, so sight along all its sides before you select the one to use and reject any with curves. Use wood which is at least 50 × 75 mm size and place the spirit level on the narrow edge. It is easier to use if you position it halfway along the plank and strap it in place with sticky

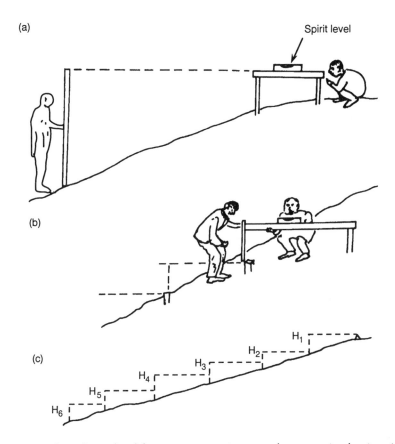

Figure 7.3 The relative level between two points may be ascertained using simple equipment such as a plank, a spirit level, some wooden pegs and a rule. In (a), two pegs have been driven into the ground so that a straight plank can be set between them. The first action is to use a spirit level to enable the pegs to be set at the same level. Once this has been completed, one person sights along the plank while the second moves downhill carrying a stake of known length (say 3 m). The stake is kept vertical and is set down at a point where the top is at the same level as the plank. The position of the stake is marked and the pegs are moved to this point. It is essential to keep clear records so that you add and subtract the correct values to calculate the height difference. A simpler method is shown in (b), where a series of pegs have been driven into the ground down a slope. The pegs have been set at a distance so that each pair may be spanned by a plank. The height difference between two pegs is measured by one end of the plank being placed on the uppermost peg and held horizontally by means of the spirit level. The height difference from the bottom of the plank to the top of the next peg is measured and carefully recorded. Do not forget to take the reading, write it down and then take it again to check it. The plank is then moved to the second peg and the procedure repeated. The height difference between the two end points is the sum of all the measurements taken as shown in (c)

tape. To level down a hillside you take the level from point to point, measuring the vertical height between the horizontal plank and the ground as you go, as shown in Figure 7.3. The total height difference between the two end points on your survey is the sum of the values at each station.

Hosepipe Method

This method uses the principle that water always finds its own level. Use a standard 15 mm diameter garden hosepipe filled with water and with a short length of clear plastic pipe stuck in each end. This will enable you to see the water level easily when you are taking the readings. This system is frequently used by builders to check the level between two walls or other parts of a new building where access is difficult for a spirit level. This method is illustrated in Figure 7.4. When you and your helper are moving the hose between readings avoid spilling the water by keeping your thumb over each end. This technique is very similar to the spirit level and plank method as the difference in height between the end points is the sum of all the height readings.

PUMPS

Having decided that you cannot obtain your water without using a pump you must look at the various types which are available and try to select one which is cheap to run. You could always use a hand pump of course, but these are very primitive and will not give you water on tap.

The majority of pumps used in small water supplies are powered by electricity. Where there is no electricity supply, a diesel engine will have to be used but these are noisy and inconvenient because you have to buy and store an adequate supply of fuel. Take care how you store the fuel so that it does not leak into your water supply. Drinking water and diesel do not make a pleasant drink! Maintenance is also more of a problem than with electric pumps as you will need to service the diesel engine whereas electrical motors need little attention.

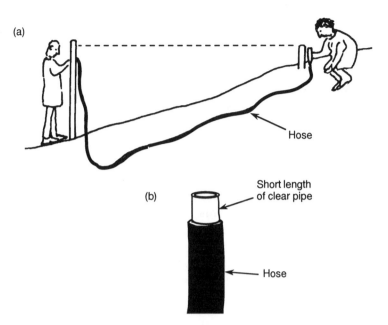

Figure 7.4 An old trick used by builders to make sure that two parts of a wall are the same height makes use of the principle that water always finds its own level. A water-filled hose is used to measure the height of one point above another as shown in (a). Use a length of flexible garden hose and insert a short length of clear plastic tube in each end (b). Fill the hose with water and then move the two ends as far apart as possible (without having a tug-of-war!). Use a measuring staff to measure the height of the first point above the second. Record the measurement and, as before, check their accuracy after you have written them down. Keep repeating the exercise as with the plank, and the difference in height between two remote points is the sum of all your measurements. You will need to keep the upper measuring point a constant height above the ground so either use pegs or a heavy object like a brick which can be moved from one measuring point to the next. When you are moving the hose, stick your thumb over the end to prevent the water being lost

The pumps that are most economical to run are those which use a free energy supply. Both wind and water power can be harnessed to drive a pump but their use declined after the early 1950s because of relatively cheap electricity. The big increase in energy costs over the last few years now makes them a very attractive option.

How Pumps Work

The action of most pumps can be divided into two parts. The first is on the intake or suction side of the pump and involves lifting water from some lower level to the pump intake. The second part of the pumping action is concerned with applying pressure to the water inside the pump to force it through the delivery pipe to wherever it is required.

A suction pump works by utilising atmospheric pressure to force water through the suction pipe. Let us consider for a moment, an open-ended tube suspended vertically in a container of water. The water level in the container and inside the tube is the same because atmospheric pressure is acting on the water both inside and outside the tube. If by some means the pressure on the water surface inside the tube is reduced below atmospheric pressure, then water will rise in the tube (see Figure 7.5). This continues until the weight of the water column inside the tube equals the difference between atmospheric pressure acting on the water outside the tube and the reduced pressure on the water inside it. The water will rise to its

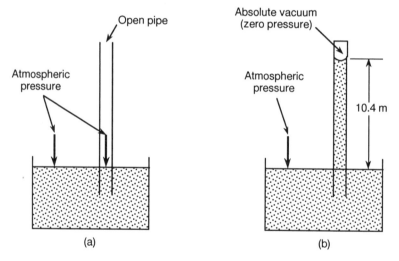

Figure 7.5 Suction pumps use atmospheric pressure to help move water. This diagram illustrates how much potential lift is provided by atmospheric pressure. In (a), an open pipe has been suspended with one end in a tank of water. The water level inside the pipe is the same as that in the tank. If the end of the pipe is then sealed and a vacuum created in the pipe, the water will rise forced up by the weight of the atmosphere on the water surface (b). This did not happen when the pipe was open as the pressure from the atmosphere was the same both inside and outside the pipe. Normal atmospheric pressure is sufficient to drive the water 10.3 m (33 feet) up the pipe in an absolute vacuum. It is not possible for ordinary pumps to create such a complete vacuum, so they cannot suck water up to quite this height. At best they manage about 7–8 m with 4–5 m being typical

maximum height when the pressure in the tube is reduced to zero, that is when a total vacuum has been produced. The weight of the water column will then be equal to that of the atmosphere.

At sea-level, atmospheric pressure is equal to a column of water about 10.4 m high. This is the height to which the water will rise in the tube at sea-level. Where the experiment is carried out at higher altitudes the water will not rise so far up the tube. In theory, therefore, a pump which can produce a complete vacuum in its suction pipe should be capable of lifting water about 10.4 m at sea-level. In practice this is never achieved as pumps are less than 100% efficient and cannot produce a perfect vacuum.

The best-designed pumps can usually achieve a lift of 7 to 8 m at sea-level but the lift of an average suction pump varies from about 4.5 to 5.5 m. This means that if you want to pump from a well or borehole when the depth of the water is more than about 5 m, some means must be found of lowering the pump into the well. This is usually done with the pump being completely submerged under water. There is almost no limit to the height to which pumps can force water out of the delivery side, so long as they have a powerful motor. Submersible pumps are designed to operate with a short intake but a long delivery section.

The limitation to suction lift is used to classify pumps into surface (or suction) pumps and deep well (or submersible) pumps. Another very common classification of pumps, however, divides them into two main types based on the mechanical principles involved. These are constant displacement and variable displacement pumps. Both types can be used as surface suction pumps or deep well submersible pumps.

Pumps move water partly by sucking through the suction intake, and partly by pushing water out through the delivery. In deciding on the size of pump you need, you will have to take account of the total head which is the height difference between the pumping water level and the highest point where you want to pump. Figure 7.6 illustrates the various aspects of head which you will have to think about.

Hydraulic Rams

Strictly speaking, a hydraulic ram is not a pump but it certainly is a very useful device for raising water. It uses the momentum of a

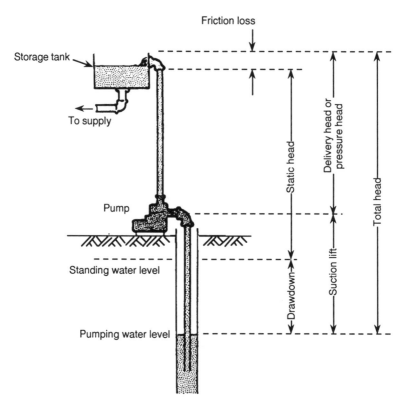

Figure 7.6 In a simple pumping arrangement there are a bewildering number of ways to look at the head differences. This drawing illustrates the various terms used. A surface suction pump is drawing water from a well and pumping it up to a storage tank. The water level in the well before pumping started is termed the *standing water level*. Once pumping has stabilised the new water level is the *pumping water level* and the difference between the two is the *drawdown*. The difference between the pumping water level and the pump suction is called the *suction lift* and the height above the pump suction that the pump is raising the water is termed the *delivery head* or *pressure head*. This is not equal to the height of the water in the tank as there is *friction loss* in the pipe which adds to the *total head* in the system. The total head is equivalent to the difference between the pumping water level and the height to which water is raised. The *static head* is the difference between the water level in the storage tank and the standing water level

relatively large flow of water under a small head to raise a small quantity of water against a large head. One theory is that it was developed in Bristol, England by a plumber who was trying to solve a problem of water hammer in the pipes of a large house. He fitted an air chamber to help absorb the shock and a relief pipe leading to a water tank. He discovered that each time a tap was closed off water would squirt into the tank and by experiment he found that the water flow could reach a tank at the top of the house.

Figure 7.7 shows the internal workings of a hydraulic ram, with the photograph in Figure 7.8 giving you a better idea of what they look like. Water from the supply stream or spring is fed into the ram through the injection pipe and flows out through the pulse valve.

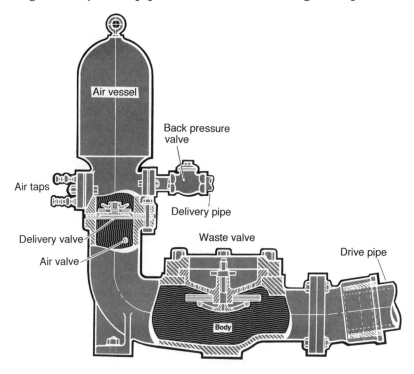

Figure 7.7 The driving flow of water enters the drive pipe seen at the right of this drawing of a hydraulic ram. The mushroom-shaped pulse or waste valve is shown open; however, the increasing pressure of the flowing water will gradually close it, with a seal formed by a rubber disc. The water flow continues to produce a pressure surge which opens the delivery valve and allows water to flow into the air vessel. The increasing pressure displaces water into the delivery pipe producing the pumping action. The air is compressed by the inflowing water, absorbing the momentum of the water producing the characteristic soft thump. The delivery valve remains open until the water in the drive pipe has almost completely slowed down and the pressure in the pump drops below the delivery pressure. The delivery valve then closes to stop any flow back into the drive pipe. A pressure drop is produced in the pump body which allows the waste valve to open and at the same time a small quantity of air is sucked in through the air valve. This small pocket of air sits below the delivery valve until it enters with the next pulse of water to replace air lost as it dissolves in the water. The cycle is then repeated producing a continuous flow through the delivery pipe. (Redrawn with permission from material supplied by John Blake (Allspeeds Limited), Accrington, Lancashire)

Figure 7.8 A hydraulic ram with a 75 mm inlet pipe and a 25 mm delivery pipe. It is capable of delivering up to 137 l/min and raising water to a height of 120 m. The main features described in Figure 7.7 can be seen, including the drive pipe on the left of the photograph, the waste valve in the centre, the delivery pipe on the far right and the air vessel at the top. (Photograph courtesy of John Blake (Allspeeds Limited), Accrington, Lancashire)

This valve is shaped rather like a mushroom and is arranged to open downwards with its own weight. As the flow of water through the ram increases, it gradually shuts the valve. Closing this valve causes an instant pressure rise in the chamber which opens a second valve, the delivery or inner valve. Water can then flow through the delivery valve into the delivery pipe. As this happens, the water pressure in the ram falls until it is too low to keep the delivery valve open. The valve then closes and stops water draining back out of the delivery pipe. The fall in pressure in the main chamber allows the pulse valve to open once more and the flow starts again through the outlet. The flow gradually increases, which in turn closes the pulse valve again and repeats the cycle. The ram is fitted with an air vessel on the delivery side. The air it contains can be compressed and so acts rather like a cushion, thereby preventing damage to the system by the shock of the pumping action.

There is a second type of hydraulic ram which does not use the supply water for the motive power. Water flowing in a polluted stream can be used to drive the ram and is prevented from coming into contact with the supply water. In this type of ram, water flowing through the chamber activates a plunger and the movement of this plunger is used to pump the clean water. The main differ-ence between the two types of ram is that the indirect system requires two intakes; one for the flow of water to power the device and the second a suction intake similar to that on a centrifugal pump, which takes water from the clean source of supply.

The water used to drive a hydraulic ram should be passed through screens to prevent leaves and other rubbish which could cause a blockage from reaching the inlet. Both types of ram will stop working if foreign material prevents either valve from closing. Hydraulic rams should be inspected regularly and cleaned out when necessary. Maintenance is very easy however, and cheap too! You are likely to find that you will only have to replace the washers every year or two, with no other spare parts to buy.

Hydraulic rams can be made to work with a driving head as low as 0.5 m of water. For efficient pumping, however, the practical minimum head is usually 2 m. A standard type of hydraulic ram can be used to deliver water to heights of up to 60 m. Greater lifts can be obtained but they need specially strengthened apparatus. Figure 7.9 shows the general principles of hydraulic ram installa-tions and the relationship between the height and delivery rate.

Hydraulic rams must operate continuously and so a pumping rate should be selected which gives an adequate flow into the storage tank. The flow into the tank cannot be controlled by a float valve without damaging the ram. Any water which is surplus to require-ments should be allowed to run to waste and the overflow must be large enough to cope.

Before starting a ram pump, carry out a few checks. See that the bolts which hold the pump together and those which anchor it down are all tight. Make sure that the drain away from the ram is clear and that the water feed is steady, having checked on the condition of the

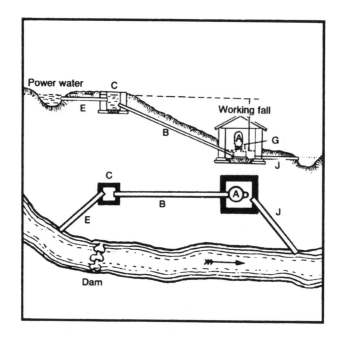

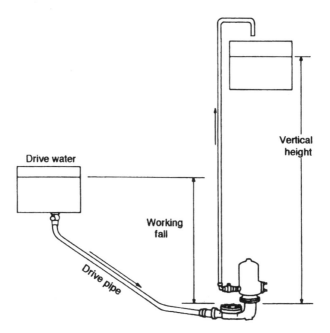

Figure 7.9

header tank. Check that the impulse valve will move and that the air snifter valve is clear of any blockage.

To start the ram, slowly open the outlet valve and then quickly open the drive pipe valve to allow a sudden inflow of water. Water should then flow through the inlet valve until it suddenly closes. If the valve then opens automatically the pump should continue to run on its own. If this does not happen you may have to help things along by pushing down on the inlet valve to re-open it. Water will then flow through it at an increasing rate until the valve shuts by itself. At this point, push down again to re-open the valve once more. If the valve does not start to operate automatically at this point you will need to keep helping it for a while longer. If the pump will not start, remove the delivery valve and check that you fitted it correctly. Check around the ram for leaks and back along the delivery pipe for both leaks and air locks, which will be suffi-cient to stop the ram from starting up. To stop the ram, simply hold the impulse valve closed or stop the water flow to the pump.

The pumping rate is controlled by adjusting the length of stroke of both the pulse valve and the delivery valve. The delivery valve is fitted with screw adjustors but the pulse valve is adjusted using a number of loose washers on the piston arm. The ram may also be fitted with an air valve to top up the air vessel. This is necessary as air is gradually lost by being dissolved in the water. Simply open the air valve when the ram is not operating and allow air to re-fill the chamber. Tighten the valve again and restart the ram.

Figure 7.9 The basic requirements for a hydraulic ram installation are a continuous supply of running water and that a driving head or working fall of at least 1 m can be achieved. It is also necessary for the driving fall not to exceed one-third of the vertical height or delivery head and the site must allow for the waste water to drain away from the ram to a point below the waste valve. The diagram shows a hydraulic ram (A) fed with water through the drive pipe (B). The driving head is controlled by a break-pressure tank (C) (see Figure 7.25), which is fed from a stream through pipe E. The water supply is pumped along the delivery pipe G and the excess water flows back to the stream along pipe J. The performance of a hydraulic ram is determined by the working fall which provides the driving energy and by the vertical height to which the pumped water is to be raised as shown in the diagram. Manufacturers will provide information on their products which will allow you to determine the size of the ram you require using the information on the required supply rate, the available working fall and the vertical height for the delivery. Where large quantities are needed or the vertical height is large, several rams may be used to provide the required supply. (Reproduced with permission from material supplied by John Blake (Allspeeds Limited), Accrington, Lancashire)

Water Turbine Pumps

There are a number of small pumps on the market which utilise the flow of a stream to pump water up to a storage tank. The flowing water rotates a turbine which in turn drives a small rotary pump. These pumps are usually installed in a small stream but have the disadvantage that at high flows they no longer work. The other disadvantage, when compared with a hydraulic ram, is that there are more moving parts and so more maintenance is required. The main advantage compared with hydraulic rams, is that they are very easily installed and so can be used on a temporary basis at one site before being moved to another.

Constant Displacement Pumps

These pumps are designed to deliver more or less the same quantity of water for a wide range of head. The power required to drive the pumps, however, is controlled by the height to which water is pumped. For example, the power supply must be doubled if the pumping or pressure head is doubled. The three main designs of this type of pump are reciprocating piston pumps, rotary pumps and helical rotor pumps.

1. *Reciprocating piston pumps* are the most common type of constant displacement pumps, a good example being a hand pump. They use the up and down or forward and backward movement of a piston (or plunger) to displace water in a cylinder. The basic principles and steps in the operation of a single acting piston pump are shown in Figure 7.10.

Piston pumps are very flexible in their use as the pumping rate can be varied by simply changing the number of strokes of the piston each minute. Other advantages are the relatively low initial cost, robust construction and ease of maintenance, which is usually restricted to the replacement of piston washers. If the piston valve or the foot valve leaks, the water column in the pump will slowly fall when the pump is not in use. In these circumstances you may not be able to get a flow without priming the pump. To do this, pour water down the suction chamber to give a water column all the way to the top.

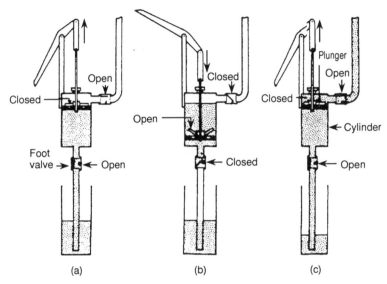

Figure 7.10 This series of drawings shows the operation of a single-acting piston pump. On the forward stroke (a), the upward moving plunger creates a suction which opens the foot valve and allows water to flow into the cylinder. The reverse stroke (b) increases pressure behind the piston in the cylinder, closing the foot valve. This opens the valve in the piston, permitting water through to the discharge side of the piston. A second forward stroke (c) pushes water from the cylinder through an open valve into the discharge pipe. At the same time, the foot valve is opened again and more water is drawn into the cylinder. Repetition of the forward and reverse strokes results in a steady flow of water out of the discharge pipe. Some pumps have the valves arranged so that water can be pumped on both the forward and reverse strokes, and are called double-acting piston pumps. The pumping action requires the section above the foot valve to be full of water to create sufficient suction to open the valve and draw in the water. Sometimes the seal round the foot valve allows a slow leak to occur. It is then necessary to prime the pump by partly filling this section with water and so enable the first forward stroke to produce enough suction to start the pumping action

Early hand pumps were made of wood by the village carpenter. Figure 7.11 shows you how it was done, but you will need excellent wood-working skills to make one for yourself. Once cast iron became common it was used to make a high proportion of hand pumps. The simplest form is called the jack pump which has a relatively short pump cylinder as you can see from Figure 7.12. Jack pumps were also made from lead, gunmetal and brass, and tended to be in common domestic use in the kitchen or scullery, and around the farm and in the dairy pumping water from cisterns and shallow wells.

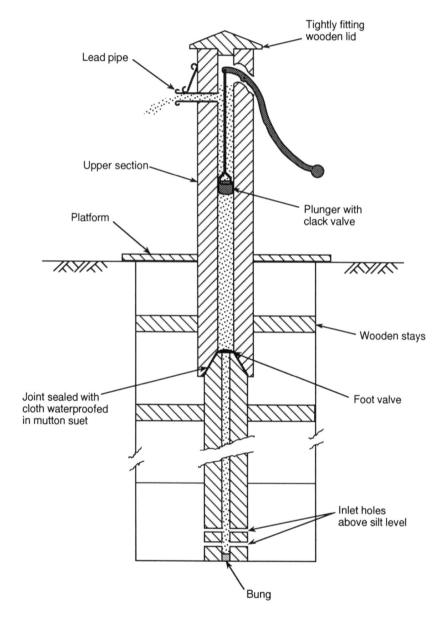

Figure 7.11

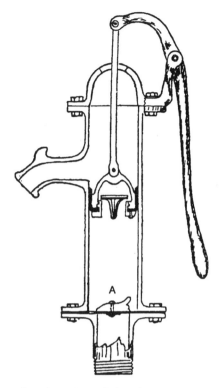

Figure 7.12 A mass-produced version of the piston pump was made from cast-iron and termed the jack pump. The bottom valve (A) comprises a leather disc cut out to form a washer and incorporating a hinged flap (or clack). The plunger has leather seals and incorporates a tapered valve as shown and is connected to the handle by a short rod. The pump is primed through the hole at the top which is made to take the rod

Figure 7.11 Early pumps were made largely out of wood. The pump body was made in two sections with the lower part made to fit into the bottom of the upper section as shown. Elm was the usual wood employed for the body and great skill was required to bore out the central hole. The auger used would be about 4.5 m long so to achieve greater lengths the hole would be bored from each end. The foot valve, made of wood and leather was positioned on top of the lower section. Tradition says that the joint between the two sections was sealed with a cloth made waterproof using hot mutton suet. The pump was made long enough to rest on the well floor so the bottom was plugged and small inlet holes were provided some distance above the base to avoid the silt accumulated on the well floor. Wooden stays were used to keep the pump in position with the main support provided at the top by the platform. A wooden plunger was made to fit the inside of the upper section, attached to the handle by a rod. The pump works as described in Figure 7.10. When necessary the pump is primed by removing the tight-fitting top. (Redrawn by permission of Cambridge University Press)

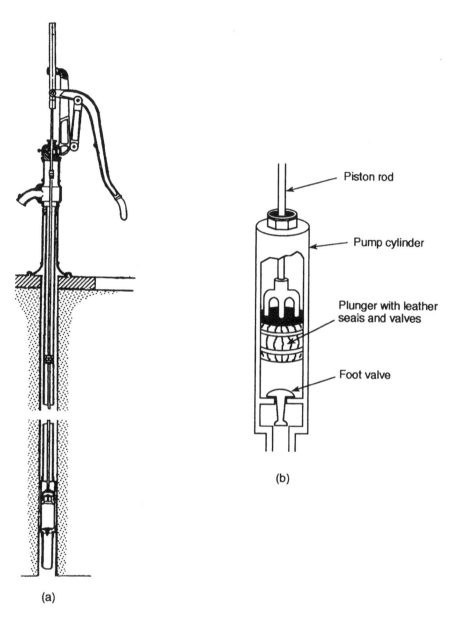

(a)

(b)

Piston rod

Pump cylinder

Plunger with leather
seals and valves

Foot valve

Figure 7.13   A deep-well hand pump works in the same way as shown in Figure 7.10. The pump body is installed in a borehole (a) with the plunger in a barrel (b) at the bottom. The barrel should be installed within 6 m of the water level

In deep wells the pump needs to have the plunger set down the well so that most of the lift is achieved by pumping rather than suction as shown in Figure 7.13. In some of these pumps the hand operation is replaced by wind power or a surface mounted diesel engine.

2. *Rotary pumps* use a system of vanes rotating in an airtight jacket to create a suction on the inlet side. This draws in water and forces it out in a continuous stream from the discharge side (see Figure 7.14). This type of pumping action is usually restricted to surface pumps.

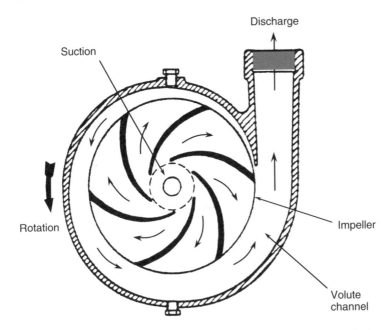

Figure 7.14 This diagram shows the workings of a centrifugal pump, which is a type of rotary pump. All rotary pumps have a rotating part usually called the impeller, which is mounted on a shaft and enclosed in a stationary casing. In order to operate, the casing and entire suction must be filled with water, so a foot valve at the bottom of the suction is essential and priming may also be necessary. In a centrifugal pump, the casing is designed so that the water is led directly to the centre of the impeller, which as it rotates at high speed, forces the water outwards towards the curved passageway known as the volute channel which leads to the pump discharge. The water moving outwards creates a reduction of pressure at the centre and this partial vacuum allows water to be drawn up through the pump suction under the influence of the atmospheric pressure acting on the water surface (see Figure 7.5). The discharge pressure developed by a centrifugal pump depends on the diameter of the impeller and the speed of rotation. The pumping rate depends on the size of the impeller and the passageways in the pump, as well as the speed of rotation. Pumps with one impeller are known as *single stage* and those with several impellers are *multi-stage* pumps

They are as limited as any pump on the height they can lift but the most efficient ones can develop a suction lift of more than 7 m provided that they have been fitted with a foot valve. Such pumps can lift water on the delivery side to heights over 20 m.

Rotary pumps are easy to make and are therefore relatively cheap. One of the main disadvantages, however, is that any sand or grit entering the pump can cause considerable wear on the closely fitting vanes, so causing a big reduction in efficiency.

3. *Helical rotor pumps* are a modified type of rotary pump. Water is driven by a highly polished metal *rotor* or screw. The rotor is a spiral shaped rod, usually made of stainless steel which rotates inside a flexible tube that is usually made of rubber and is called a *stator*. The rotor has flexible mountings so that it can rotate eccentrically within the stator as shown in Figure 7.15. This pushes a continuous stream of water forward along the cavities of the stator. The water acts as a lubricant between the two parts of the pump and improves the efficiency.

These pumps can be used both at the surface and in a deep well and are usually driven by either engines or electric motors. It is also possible to rotate the drive shaft by hand or to use wind power although both these methods give a reduced pumping rate. The main advantage of this type of pump is that sand or other suspended material in the water can be sucked through the rotor without wearing it out.

Variable Displacement Pumps

In this general class of pump, the pumping rate is altered by changes in the head. An increasing head reduces the pumping rate. The two main types of variable displacement pumps used in small water supplies are centrifugal and jet pumps.

1. *Centrifugal pumps* are the most common type of pump in general use. They operate by a rotating impeller which is designed so that centrifugal force is used to pump water. There are many designs of this sort of pump and they are used as both surface suction pumps and deep well pumps. The rotary action to drive the turbines of a deep well pump can be delivered from a surface motor through

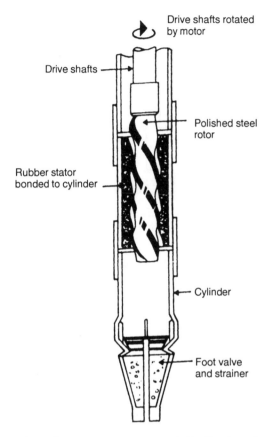

Drive shafts rotated by motor

Drive shafts

Polished steel rotor

Rubber stator bonded to cylinder

Cylinder

Foot valve and strainer

Figure 7.15 A version of a rotary pump uses a helically shaped rod or *rotor* as the impeller. The rotor is usually made from polished hard chrome plated stainless steel and is positioned inside a moulded rubber cylinder or *stator* which is bonded to the pump body. The rotor and stator have a similar shape, so that during rotation a series of sealed capsules is produced between the two components, which progresses water through the pump. The diagram shows a version suitable for use in a well. (Redrawn with permission from H₂OWaste-Tec, Stockport)

rotating shafts. Electrical submersible pumps are much more common and have an electrical motor which is contained within the pump assembly in a sealed chamber. An example of an electrical submersible deep well pump is shown in Figure 7.16.

When water has to be pumped against a high head using a centrifugal pump, several stages are used. Each stage is, in effect, a mini-pump and contains an impeller. This multi-stage design is used in both surface and deep well pumps but is especially common in

To surface

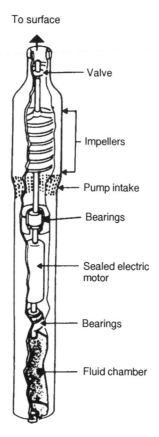

Valve

Impellers

Pump intake

Bearings

Sealed electric
motor

Bearings

Fluid chamber

Figure 7.16 An electrical submersible pump is a type of centrifugal pump made so that it will operate in a narrow space. The pump is powered by a sealed electric motor which rotates the impellers. The motor is positioned at the bottom of the pump with the intake immediately above it. This ensures that water is drawn continuously past the motor to provide a cooling action. Generally several impellers are used to provide a sufficient lift to raise the water to the ground surface and into the water supply system. In some instances the pressure produced by the submersible pump is used to maintain a working pressure in a pressure tank sufficient to move the water through the distribution system (see Figure 7.23)

deep well pumps made for use under high lift conditions. The number of stages in a particular pump does not significantly alter the rate of pumping but the pressure head developed is increased in direct proportion to the number of stages used. If each stage of a four-stage pump develops a pressure of 15 m head of water, for example, that developed by the multi-stage pump would be 60 m head of water.

2. *Jet pumps* are limited to use in wells or boreholes and have the advantage of being fairly cheap to buy and maintain and are in fairly common use in private water supply systems around the world. They are really an adaptation of a surface centrifugal pump where part of the pumped water is returned down the well and into the pumping main under pressure. The pipework is designed so that the water passes into the pumping main through a tapered nozzle. As the water flows through this nozzle its velocity suddenly

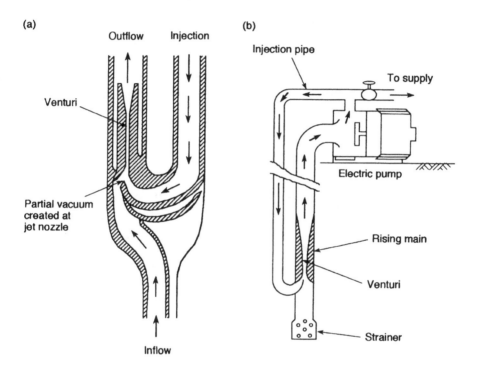

Figure 7.17 Jet pumps are used to pump a wide range of liquids. In small water supplies, however, they are most frequently used in shallow wells and boreholes. A jet pump works by generating a partial vacuum using the flow of water through specially designed pipes. Water is injected down a small diameter pipe (typically 25 mm) into the jet nozzle (a). This conducts the injected water through a short curved length of pipe so that the water is squirted into the mouth of a Venturi tube facing up the borehole. The Venturi tube has a narrow bore which widens at one end. As the water flows into this wider section a partial vacuum is produced which sucks water in through the inlet. As the injection is a continuous operation, a pumping action is established which forces more water up the pipes than flows down. Consequently, the outflow pipe has a larger diameter (typically 38 mm). The pumping action is usually provided by a surface-mounted, electrical rotary pump (b)

increases and causes a reduction in pressure. This causes a partial vacuum which sucks water from the well through the intake pipe into the delivery main. The arrangement for jet pump installation is shown in Figure 7.17.

Unfortunately, they are inefficient and use more power than any other type of deep well pump because of the need to re-circulate part of the flow. In fact, as the water level falls the proportion of water being passed back down the borehole increases significantly and so less water is obtained. Although jet pumps are generally inefficient they have a number of desirable features which make their use popular in some small domestic water supply installations. Their advantages include adaptability for use in small wells down to 50 mm in diameter, low purchase price and maintenance costs, simplicity of operation, and accessibility of moving parts which are all above the ground surface. Jet pumps can be noisy so make sure that you do not install one inside your house.

Wind Pumps

A common sight in rural areas are the multiple-vaned windmills which are used to pump water out of wells and boreholes. These wind-vanes are fitted with an automatic governing device and are usually installed on steel-framed towers so that they are between 15 and 20 m above the ground. This is to make sure that the windmill is above surrounding obstacles in order to provide a clear sweep of the wind to the vane. Windmills require a wind speed of at least 10 km/h to turn them, although wind speeds in the order of 25–40 km/h are needed for optimum operation. They are traditionally used to drive reciprocating pumps (see Figure 7.18), but there is no reason why the rotating shaft cannot be connected to the rotary drive shaft on a helical rotor pump or a rod-driven deep well turbine pump.

When designing a supply based on wind power pumping, it is important to remember that the wind will not blow all the time. On average, a wind pump will perhaps be in operation for some 8 h out of every 24 h so it is essential to provide a storage tank equivalent to some four to seven days' water use. As an alternative, have a back-up system so that the pump can be driven by a small diesel engine or even by hand.

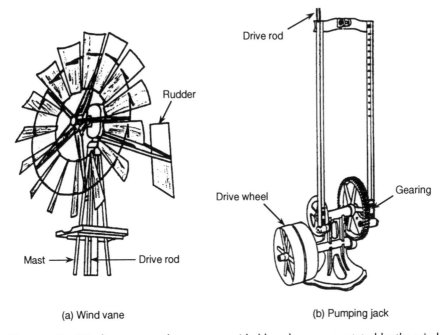

Drive rod

Rudder

Drive wheel

Gearing

Mast

Drive rod

(a) Wind vane

(b) Pumping jack

Figure 7.18   Wind pumps use the power provided by a large vane rotated by the wind to power a pump of the type shown in (a). Modern wind pumps are designed to be self-regulating to eliminate excessive speeds which might put undue strain on the pump or mechanical equipment. They are also designed so that they automatically move the position of the wheel that is facing the wind. If the full area of the wheel face is exposed to the full force of the wind there is danger of it being damaged and the tower being blown over. To operate satisfactorily the wind vane must be positioned on a tower of sufficient strength and height. Tall buildings and trees should be avoided as these may cause eddies and an uneven flow of wind. As a rule of thumb, the tower should be about 3 m higher than the tallest trees or buildings within 100 m of the well. The rotation of the wind vane is transferred to an up-and-down action through a series of gears. Rods moving up and down then transfer the motion to a gearing mechanism (or pump jack as shown in (b) which transfers it on to the pump rods. The pump is usually a deep-well piston pump similar to the one shown in Figure 7.13. The pump jack has a wheel so that it may be driven by a motor if necessary using a drive-belt

PIPES

You will need a network of pipes to distribute water to your house and to the various buildings and other points where water is required. A bad design or a poorly built pipe layout will result in an inadequate water supply however much effort you have spent with the source, pump selection or the size of the storage tank. Care is

Table 7.1 Friction loss factors for HDP plastic water pipe are shown in this table expressed in m/100 m (i.e. %) and apply to gravity flow through pipes. There are slight variations in pipe specifications between manufacturers which results in differences in flow characteristics, so the values given in this table must be regarded as approximate. Losses of less than 20% are regarded as negligible. The values given are up to the maximum recommended flow velocity of 3.0 m/sec. Factors with flow velocities less than 0.7 m/sec are marked with an asterisk. Class III and Class IV refer to the strength of the pipe wall. (Reproduced by permission of UNICEF Kathmandu from *A Handbook of Gravity-Flow Water Systems* by T. D. Jordan, published by Intermediate Technology Publications Limited).

Flow	Class III (6 kg/cm²)				Class IV (10 kg/cm²)				
	32 mm	50 mm	63 mm	90 mm	20 mm	32 mm	50 mm	63 mm	90 mm
0.10	0.22*				3.1*	0.31*			
0.12	0.30*				4.5*	0.40*			
0.14	0.38*				5.6*	0.52*			
0.16	0.48*				7.3	0.67*			
0.18	0.58*				9.0	0.81*			
0.20	0.72*				10.6	0.99*			
0.225	0.87*				13.4	1.22*			
0.25	1.08*				15.7	1.43*			
0.275	1.27*				18.5	1.74*			
0.30	1.42*				21.8	2.02*			
0.35	1.88*	0.22*			28.0	2.71*	0.25*		
0.40	2.44	0.28*			36.2	3.36	0.40*		
0.45	2.87	0.34*			45	4.0	0.49*		
0.50	3.70	0.40*			54	4.9	0.59*		
0.55	4.1	0.47*			63	5.7	0.71*		
0.60	4.9	0.56*			73	6.7	0.81*	0.27*	
0.65	5.6	0.63*				7.8	0.94*	0.31*	
0.675	5.9	0.67*				8.4	1.00*	0.34*	
0.70	6.3	0.72*	0.25*			8.8	1.06*	0.36*	
0.75	7.3	0.81*	0.28*			10.1	1.23*	0.40*	
0.80	8.2	0.90*	0.31*			11.2	1.34*	0.45*	
0.90	10.0	1.11*	0.37*			13.4	1.67*	0.54	
1.00	11.9	1.34*	0.45*			16.4	2.00*	0.66*	
1.10	14.1	1.57	0.54*			19.8	2.37	0.78*	
1.20	16.5	1.90	0.63*			22.6	2.77	0.92*	
1.30	19.0	2.18	0.73*			26.4	3.19	1.08*	
1.40	21.6	2.46	0.82*			30.2	3.61	1.23*	0.22*
1.60	27.4	3.05	1.03*			37.5	4.5	1.52	0.28*
1.80	33.6	3.81	1.30	0.24*			5.6	1.85	0.35*
2.00		4.6	1.55	0.28*			6.7	2.24	0.41*
2.20		5.5	2.46	0.32*			8.3	2.69	0.49*
2.50		6.7	2.24	0.40*			10.1	3.25	0.60*
2.70		8.3	2.74	0.49*			12.2	3.92	0.73*
3.00		9.5	3.08	0.56*			13.9	4.6	0.84
3.20		11.2	3.77	0.67*			17.4	5.5	0.99
3.50		12.6	4.1	0.74			18.9	6.2	1.11
3.70		14.4	4.8	0.84			21.3	6.9	1.29
4.00		15.7	5.4	0.96			23.7	7.7	1.43
4.20		18.1	6.0	1.04				8.7	1.57
4.50		20.1	6.6	1.06				9.5	1.78
4.70			7.4	1.31				10.8	1.97
5.00			7.8	1.41				11.8	2.13
5.50			9.4	1.70				13.4	2.46
6.00			11.1	2.00				15.9	2.91
6.50			12.3	2.24					3.36
7.00			14.6	2.97					3.89

20 mm HDP is only available in the Class IV series.

needed in choosing the right pipe for the job, taking account of the pipe material, its size and how much pressure it needs to withstand.

Water Flow in Pipes

The theories which cover water flow in pipes are complicated and the comments and suggestions made here are simplifications which, nevertheless, are likely to work satisfactorily in most cases. The factors which control the quantities flowing through a pipe are the *head* (i.e. the difference in level between each end), the pipe size, the material used to make the pipe and complications in the pipe runs such as curves, bends, branches and joints. The last two factors relate to the *friction losses*.

Just as the friction on a ball rolling across a football field will slow it down, the friction generated by the water flowing against the pipe walls reduces the flow. The head is often increased to get round the problem of a reduced throughput. The amount of this increase is equal to the friction losses or head losses in the system. Head losses are increased by joints, bends, tee-pieces and branches. As a result, a greater driving force is needed in complicated systems. One way round the problem is to use pipe which is the next size up to the one needed for an equivalent flow in a simple system.

Table 7.1 shows the friction losses in pipes with different diameters and at different flow rates for standard lengths of pipe. To work out the losses in your system simply multiply the total length of pipe in your system, or in any section you are considering separately, by the appropriate factor in the table.

Include friction losses caused by pipe fittings and valves in your calculation by converting the friction of each fitting to an equivalent length of pipe. Table 7.2 gives the conversion factors you will need for a range of fittings. To work out the friction loss, multiply the table factor by the pipe diameter and the result is the equivalent length of pipe in the same length unit as the diameter. As pipe diameters are usually in inches or millimetres you will have to convert to feet or metres as appropriate. Calculate the pipe length equivalent for each fitting, then add this to the length of pipe you will use, and use Table 7.1 to give you the friction loss in feet or metres head of water. This value is the minimum height that your storage tank should be above the highest water outlet in the system.

Table 7.2 Friction loss in pipe fittings expressed as an equivalent straight length of pipe

Bends	
short radius	30–35
round elbow	45
medium radius	15
close return	100
Tee-pieces	
straight through	10
side outlet	50–55
swept tee	20–25
Sudden enlargement	45
Taper	45

The equivalent length is the table factor multiplied by the pipe diameter and expressed in the same units as the diameter. For example, a round elbow bend on a pipe diameter of 50 mm has the equivalent friction loss as a straight length of 50 mm diameter pipe 45×50 mm $= 2.25$ m long.

If you are using a pressure tank calculate the pressure needed by adding the height of the upper outlet above the tank and the friction loss value.

Planning the Layout

It is usual to have the water supply enter a house through the kitchen so that the drinking water can be drawn off first. After that the water is distributed round the house, feeding other cold water outlets to hand basins and lavatories, and also supplying the hot water system. When you are planning the points in your house to provide water, do not forget an outside tap. It is useful for watering the garden, washing down the yard and washing your car. If you have a large garden it maybe convenient to locate an extra tap near the garden shed. The plumbing system inside your house is covered by a large number of DIY manuals and is beyond the scope of this book.

If you are designing the pipe network for a farm or smallholding, decide first of all where water supplies will be needed. It is important to have a supply point in all the buildings where animals and poultry are kept and in the dairy or milking parlour for washing down, cleaning the cows' udders and milking equipment and for cooling the milk. Provide a good supply in the farmyard for hosing down and do not miss out a fire hydrant! Make sure that you locate it at least 50 feet (15 m) away from all buildings. During a fire, heat or smoke may prevent access to water points fixed to the walls of a burning building. An alternative is to build a duck pond at a convenient location to provide an emergency store of water for fire fighting.

Your pipe network will run from the source to the storage tank or reservoir, and then on to the house and various buildings and places where water is required, and is likely to involve a few joints, branches and bends. At each point you need to work out the water flow needed to meet the peak demand. This will determine the size of the pipe in each part of the network.

The easiest way to tackle the job is to make a sketch of your farm layout and identify the places where water is needed, as in Figure 7.19. Look at the largest peak water demand in each building as you did when thinking about your pump; it may be the washing machine in the house and the drinking troughs in the cowshed. Do not make the mistake of adding up all the peak demand flows or your main pipe from the source will be much bigger than necessary. It is most unlikely that all the biggest uses will require water at the same time. Your common sense will tell you which are likely and help you determine the maximum flow needed in each part of the pipe network.

The next job is to decide where the pipes should run. It is a good idea to keep the pipe runs as short as possible, so go in straight lines between points when you can. Try to make the gradient even. In hilly areas this may mean not going in a straight line, but instead zig-zagging down the hillside in much the same way that you would plan a road. Try to avoid digging up roads and paved area. Besides adding to the work of pipe laying, it will make it more difficult in the future if you need to expose the pipe to mend a leak. Incorporate a few control valves or stop-cocks into the pipe system so that you will be able to isolate sections of pipe for maintenance or in the event of a burst.

Work out the pressures in your system from the different levels of your storage tank and outlet points. Calculate the height difference in metres between the water level in the storage tank and the

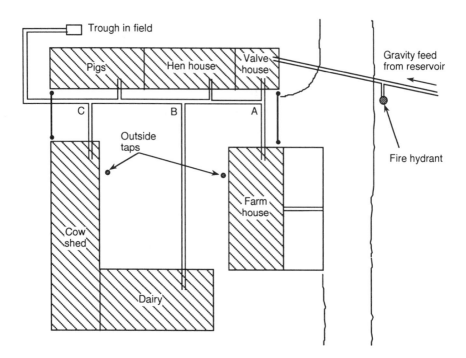

Figure 7.19 The first step in planning the layout of your distribution pipes is to sketch out a plan of your house or farm, identifying all the points where a water supply is needed. In this example water flows by gravity from a storage reservoir to the farmhouse and buildings. The main feed pipe goes to a small out-house where the main valve and flow meters are installed. The distribution network leaves the valve house and runs along the side of the buildings with junctions for the feed to the house (at A), the dairy (at B) and the cow shed (at C). The pipe sizes should be large enough to accommodate the peak demand flow for each location. Once you have worked out how much water is needed in each building use the information provided in this chapter to calculate the pipe sizes. By the time you get to the single runs to the house, dairy, etc., it is likely that you will only need a 15 mm pipe. An important feature of this system is the fire hydrant situated across the lane from the farm, which takes its supply directly from the reservoir. An alternative to providing a fire hydrant, which may be also used as a back-up supply for fire fighting, is to make a small pond. Dig out an area next to a stream to create sufficient storage for putting out the flames if your property catches fire. For the remainder of the time it can be an attractive water feature and somewhere for your ducks to swim

lowest outlet point. This will be the maximum pressure in your distribution system. The pressure at any point in the system will be the difference in level between that point and the tank water level. These pressures can be measured in feet or metres head of water, or using units of pressure such as pounds per square inch (psi). You may find it easier to work it out in feet or metres and convert to psi

(1 psi = 2.3 feet or 0.7 m head of water). Compare your calculations with the pressure rating of the pipe you are planning to use for each part of the network and only use pipe which is strong enough. Pipe pressure ratings are usually given in psi or metres head of water.

Pipe Size and Materials

Pipe sizes are given as the diameter but it is the cross-sectional area which determines how much water can flow through it. As a result, when you increase the pipe size, the flow capacity increases much more rapidly. You can use the relationship between pipe size and flow to choose the sizes needed to make up your network. Before you do that, however, you will have to decide on the pipe material because the flow capacities of a given size of pipe vary from one material to another.

The two main materials used for water distribution pipes are galvanised steel and plastic. Copper pipe is often used inside buildings but its cost makes it prohibitive for mains in general.

Galvanised steel pipe comes in standard lengths of 20 feet (6 m) and is jointed by threaded couplings. It is covered with a protective coating of zinc which resists corrosion and significantly increases its life. Nevertheless, aggressive water will attack metal pipework, thereby reducing its life and increasing the concentrations of dissolved metals in the water supply. Partly for these reasons I favour the use of plastic pipes where possible, particularly polythene (polyethylene).

There are a number of different plastics used for pipes, each one having different physical and chemical properties. Plastic pipes are light and easy to handle and are generally chemically inert. They may be squashed or broken so require protection. In selecting pipe for drinking water, make sure that it meets the local drinking water or plumbing standards. Using the wrong pipe could mean that toxic chemicals slowly leak into your drinking water.

There are five main types of plastic pipe commonly used for water pipes.

1. **Polythene**, which is also called "polyethylene", is sold in long coils which reduces the number of joints needed, with benefits in ease of installation and reduced friction losses. Polythene pipe is easily joined to steel or copper and has an in-built resistance to freezing.

It tends to be used by water companies for the smaller diameter end of their pipe networks and is coloured either black or blue. It is available in diameters up to 6 inches (300 mm) and is jointed by special compression joints. Polythene pipes are sold in different grades according to the water pressures they can withstand.

2. **uPVC** pipes are sold in standard lengths up to 20 feet (6 m). It is much more rigid than polythene but can be bent if heated gently. The pipes are normally joined by a collar which is solvent glued in place. Again it is rated according to the pressure.

3. **CPVC** is a similar plastic to uPVC but is more resistant to attack from aggressive waters. It is used in the same way as uPVC pipes.

4. **Polypropylene** pipes are similar to uPVC but tend to be used for roof drainage or waste-water systems. This plastic cannot be jointed by solvents and needs special push-fit or ring-seal joints.

5. **Polybutylene** is another rigid plastic and is used to make pipes for both hot and cold water systems. Like polypropolene, these pipes must be jointed with compression fittings.

All these pipes are available from all good class plumbers' and builders' merchants. Your supplier will advise you on the correct grade or class of pipe to buy. Buy the joints, fittings and adhesives at the same time.

The diameter of pipe you need depends on the amount of water the pipe is to carry, the material it is made from and the slope of the pipes. As a general guide, 15 mm diameter pipe made of polythene or uPVC will take flows up to 0.1 l/s. Pipes of 25 mm diameter can take 0.4 l/s and those with 50 mm diameter can carry a flow between 1 and 1.5 l/s. If the pipe has been installed at gradients in excess of 1 in 15 the flow can increase to more than 3 l/s. Use these figures as a general guide. Most needs can be satisfied with a supply of 0.4 l/s, so 25 mm diameter pipe is adequate in most cases. If greater flows are needed, look up the pipe size you need on Figure 7.20 which shows the pipe size to choose for a given flow rate. It works for a range of heads and so will apply in most cases.

Valves

Valves are an important enough part of a water supply system to warrant special attention in the pipe network design. You should

(a) plastic pipe

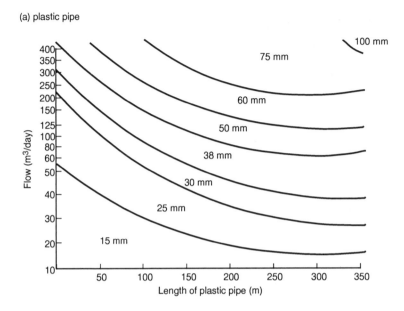

(b) galvanized steel pipe

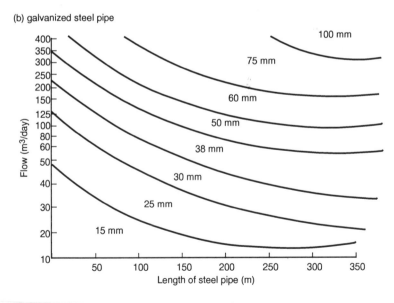

Figure 7.20 *For caption see p. 175*

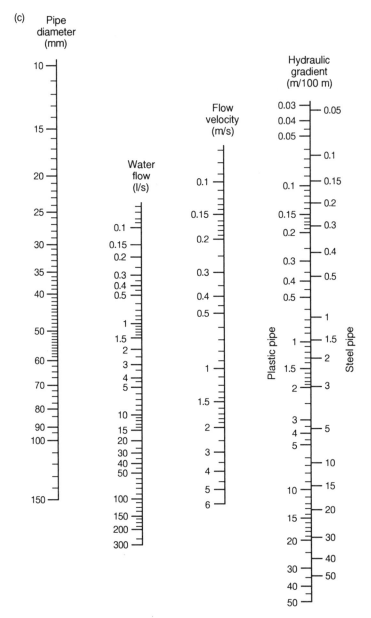

Figure 7.20 (continued)

have enough valves so that you can shut off the supply into every part of your system, including the tank or storage reservoir and all separate pipe runs. Each building which has a water supply should have its own stop valve, as should outside taps and cattle troughs. Enclose each valve in a small chamber made from pre-cast concrete, or in a length of pipe, and keep them maintained so they will work when you need them. This means turning them on and off at least twice a year and checking for leaks around the valve top.

Pipework Protection

When installing your pipes bury them in trenches deep enough to protect them from frost. The depth depends on where you live. For example, in much of England 75 cm is adequate but in colder areas, such as near the Great Lakes, bury them at least 3 m deep. Installing pipes at these depths is hard work if you are going to dig the trenches yourself. Seek the help of specialist contractors who have machines which can lay coils of polythene pipes without needing to dig a trench. Make sure that you mark the position of the buried pipes so that you can find them easily in the future.

All joints must be watertight. This is not only to prevent leakage and waste, but also to prevent polluted water getting into leaking

---

Figure 7.20 The three graphs presented here will help you to decide on the diameter of the pipe required for each part of your water supply system. Use the first graph (a) to decide the size of plastic pipe you want to use. Convert the required flow rate to cubic metres/day and use the graph to compare this flow with the length of pipe run to decide on the appropriate pipe size. If the value falls on the line between two pipe sizes, choose the larger one. If you are using galvanised steel pipe, use graph (b). You will notice that generally a larger-diameter steel pipe is required for a given flow rate and pipe run compared with plastic. This is caused by the plastic surface being smooth and having a lower frictional resistance to flow. The third graph (c) is a nonogram which relates the pipe diameter, flow rate, flow velocity and the hydraulic gradient. Simply use a ruler to join any two of the variables, and the appropriate value for the others will fall on the same straight line. For example, a flow of 0.5 l/s over a hydraulic gradient of 4 m/100 m will require a plastic pipe diameter of 25 mm. A steel pipe of 30 mm diameter would be required for the same flow and gradient. The nonogram also shows that if the hydraulic gradient is low, say, 0.5 m/100 m, and a flow of 5 l/s is required, a plastic pipe of 95 mm diameter would be needed (or 100 mm for galvanised steel). (Based on the Colebrook White formula and information provided by Durapipe, Glynwed Plastics Limited, Cannock; Hepworth Industrial Plastics Limited, Sheffield; and Wavin Industrial Products Limited, Durham)

pipes when they are not under pressure. Do not bury compression joints; locate them in small inspection chambers. These can be made from a length of salt glazed pipe or plastic drain pipe with a cover on the top.

Protect pipes which cross ditches and roads. They can either be buried at least 30 cm below the bottom of the ditch or placed in a

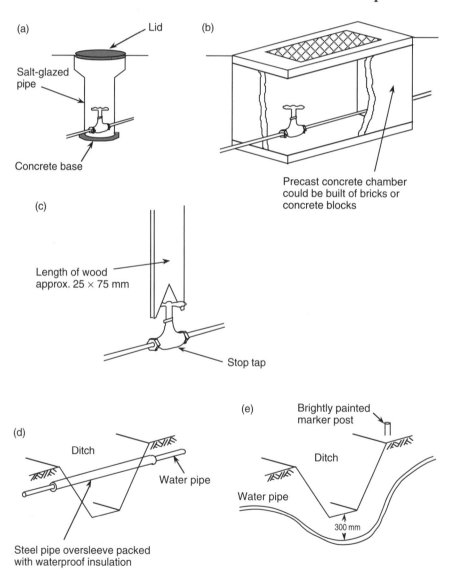

Figure 7.21

protective casing made from a length of steel pipe. Surround the water pipe with lagging to prevent winter freeze-ups. Use a steel pipe as a sleeve when plastic pipe runs under a road to prevent it being squashed by the weight of the traffic. Do not forget to have marker post to show where your pipe is buried beneath a ditch so that the pipe is not broken each time the ditch is cleaned out. Figure 7.21 shows some of the ways in which you can protect your pipework.

TANKS

A storage tank is an essential part of any water supply system. Having storage available means that you can use water without having to pump and it overcomes the problem caused by water supplies being little more than a trickle. Storage tanks fall into two categories: open tanks which feed the distribution system by gravity flow; and pressure tanks which use the supply pump to provide enough energy to force the water round the pipework.

Open Tanks

The tank should be big enough to store at least two days' supply. The bigger it is, the more easily you will be able to cope with droughts and other emergencies. The storage in the tank will also

---

Figure 7.21 Pipes which carry water around your supply system need to be protected at points where they are most vulnerable. In general, the pipes should be set in a trench deep enough to provide protection from frost or to prevent unauthorised access to it. Several features of pipelines require special attention. Protect stop taps or valves by setting them in a chamber. These can be made from a length of pipe as shown in (a); or by using a precast concrete box specially made for the job as shown in (b). If you do not have a stop tap key you can make one by cutting a notch in a length of wood as shown in (c). Select wood of a size which will be strong enough not to split if you have to apply pressure to open a stubborn tap. Two methods of crossing a ditch are shown in (d) and (e). The first method has the pipe encased in a length of steel pipe which is packed with insulating material. This approach can also be used if you have a convenient bridge and have decided to strap the pipe to the side. If you do not provide frost protection in areas liable to freezing you will be faced with lots of repairs and periodic water shortages.The alternative method of burying the pipe may provide more protection but you must mark the position of the pipe so that it is not damaged when the ditch is being cleaned out. Protect a pipe where the trench crosses a road or track by placing it in an oversleeve as in (d) and burying it at least 600 mm below the road

mean that you can shut down the water source for maintenance and have enough water for unexpectedly large water requirements. Your supply pump is controlled by a float switch, which will operate the pump for a limited number of times each day rather than turn on and off each time you turn on a tap. This will prolong the life of your pump. Figure 7.22 illustrates this type of storage tank.

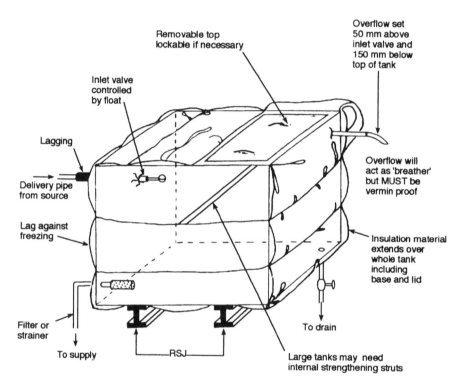

Figure 7.22   Adequate storage may be provided by a simple tank in many small water supply systems. It is important to locate the tank so that water may flow to all the points where it is needed under gravity. A tank of water is heavy, so make sure that the tank is well supported on strong beams or RSJs (a type of steel girder). The drawing shows a tank with the inflow controlled by a float valve. The outlet is set about 100 mm above the base of the tank and is fitted with a strainer. There is an overflow set about 50 mm above the inlet pipe and 150 mm below the top of the tank. The overflow is fitted with a mesh to stop vermin climbing inside the tank. A drain pipe fitted with a valve is set into the bottom of the tank. Access is gained through a tight-fitting cover which should be lockable in most cases. All the pipes and the tank itself should be insulated. Tanks are frequently made from galvanised steel but these days plastic and glass-reinforced plastic tanks are available which are made to meet water supply regulations and generally are to be preferred

Storage is not the only advantage provided by a tank. If it is positioned correctly, water can flow from your tank to all your supply points under gravity. Make sure that the tank is above the highest off-take point by a sufficient height to overcome friction losses in the pipes. In most cases where you are only thinking about the supply to your house this distance need only be about 10 feet or 3 m. Table 7.1 will give you a more detailed guide on the height you need to overcome friction losses in a larger system. The tank should be high enough so that the minimum gradient of any pipe running between the tank and any supply point is 1:50. If necessary put the tank on a tower or in a tall building to achieve this height. Tanks can also be used to reduce the pressure in a distribution system. Such tanks are called *break-pressure tanks* and are described below in the section on storage reservoirs.

Take full advantage of the modern materials which are used to make tanks. Plastic and fibreglass tanks are no cheaper than steel but their light weight makes them easier to install and maintain. Both these advantages will save you time and money.

All tanks must be adequately protected from freezing during the winter. They should also have a good top which is dustproof and can be locked. Make sure that vermin cannot get into your tank. A fairly fine mesh over all openings, such as the overflow, will prevent even flies from getting in.

If any animals or insects get into your tank the chances are they will be unable to get out again and will drown. As they decay, the water may develop an unpleasant taste and your family could get stomach disorders. One serious illness that can be caused by infected rats dying or even urinating in the water supply is Weil's Disease. This is a form of jaundice (Leptospiral Jaundice) which needs special treatment but can easily be mistaken for other illnesses. Its consequences are serious and long term, so take precautions to keep all vermin out of your water supply.

Pressure Tanks

Pressure tanks are used in many private water supply systems. A pressure tank is sealed so that as it is pumped full of water the air inside is compressed, increasing the pressure in the tank until the pump is turned off by a high-pressure control switch. When the water is drawn off, the compressed air forces the water out under pressure. As the water drains out, the pressure drops. However, the water supply is maintained by the pump which is switched back on by a low-pressure control switch. Figure 7.23 shows how these tanks are incorporated into a water supply system.

Only about 20–30% of the tank's total capacity can be used so pressure tanks do not provide much storage. Their main attraction is that they store enough water to supply small demands thereby reducing the need to pump and lessening pump wear. The sealed system also eliminates the chance of contamination problems that open tanks tend to suffer. Generally a tank is installed with a total capacity equal to ten times the volume that the pump delivers each

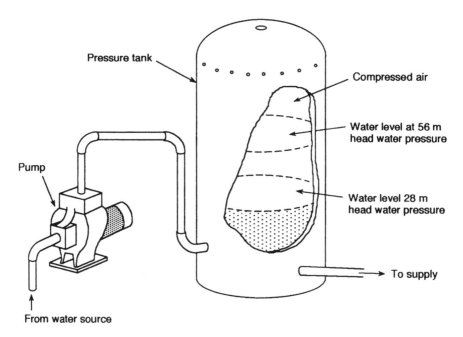

Figure 7.23 A pressure tank is installed as the starting point of the water distribution system. Water is pumped into the tank either from a surface pump (as shown) or a submersible pump. The pump is controlled by a pressure switch which operates between two pressure settings. In our example the lower pressure is 28 m head of water, and when the pressure in the tank falls to this level, the pump is switched on. The upper setting is 56 m head of water, and as the pressure in the tank reaches this value the pump is turned off. The volume of water in the tank between the two water levels represented by these pressures can be used without the pump being switched on. The tank capacity and working pressures should be selected so that this available storage is sufficient to meet typical short-term uses and so avoid excessive pump start-ups

minute. For example, a 10 gallons/minute pump requires a 100 gallon tank.

Pressure tanks often incorporate a membrane to separate the water and air and so reduce air losses from it being dissolved. In some instances the air is totally enclosed in a sort of balloon. If the air is in contact with the water and slowly dissolves, the tank will become waterlogged and will no longer operate efficiently. This problem can be overcome by topping up the amount of air in the tank through a special valve. However, when the air is separated from the water by a membrane, waterlogging is prevented.

In small systems an elastic pressure cell may be used. This is like a miniature version of a pressure tank and holds only about 10 litres. The cell consists of a metal cylinder with an elastic liner. As the water is pumped into the cell the pressure on the elastic liner increases rather like the bladder of an old-fashioned football being pumped up. When the water is needed it is forced out under the pressure provided by the elastic liner. As the storage in one of these cells is small, the pump will operate more frequently — almost each time that water is used. This disadvantage is off-set to some extent by using two or three elastic cells in series.

STORAGE RESERVOIRS

Storage reservoirs are basically big tanks so all the comments which apply to tanks apply to them also. If you are designing a water supply for a farm or a fairly large establishment like a hotel or school you will need a large volume of storage. Large tanks can be used but there are many problems in their construction. The weight of a large tank full of water means that specially strengthened supports will be needed unless it is placed directly on the ground.

Big tanks are expensive and it may be much cheaper to build a reservoir out of bricks, concrete blocks or reinforced concrete cast on site. With reinforced concrete you will need to make a frame out of steel rods fastened with wire for the reinforcing. Build wooden shuttering to hold the concrete while it is setting and use a standard 1:2:4 mix concrete to make the walls. Figure 7.24 shows the main points to consider when building a storage reservoir.

Make sure that you site the reservoir high enough for water to flow to all parts of the distribution system. Take into account the

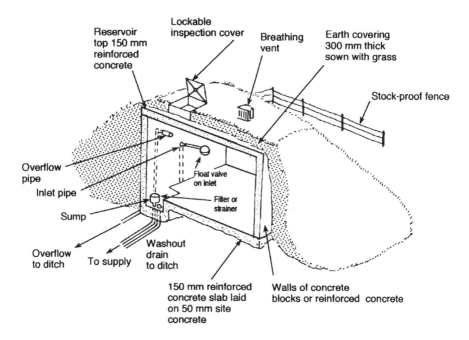

Figure 7.24 Large water supply systems where a significant storage volume is needed usually incorporate a reservoir or buried tank situated on a hillside at an elevation which will permit water to flow to all the points where it is required. A typical construction is shown here where the tank has been constructed in reinforced concrete. Alternative construction materials include concrete blocks, bricks or, where medium-sized tanks are required, glass-reinforced plastic. This last material is available in drinking water quality and is to be preferred if possible. The main features are similar to the storage tank shown in Figure 7.22. The inflow is controlled by a float valve and the outlet is set above the floor of the reservoir and fitted with a strainer. The overflow is taken to a ditch and set at 50 mm above the inlet pipe. A breathing vent is set in the centre of the reservoir and has a fine mesh to prevent insects and other small animals from finding their way in. Similarly the overflow should also be fitted with a vermin-proof mesh at its lower end. Fit valves on the inflow and outflow pipes as well as the drain or washout pipe. Access to the reservoir is through a lockable inspection cover. The tank is insulated and protected by a layer of earth at least 300 mm thick which forms sloping sides. The reservoir should be surrounded with a stock-proof fence with a locked gate

need to have extra pressure to overcome friction in the pipes. Use Table 7.1 as a guide in working out the amount you need. As with storage tanks, try to choose a site where the minimum gradient of the pipes running to the supply points is 1:50. Protect your reservoir from frost with a layer of earth at least 30 cm thick in countries like Britain and more than 1 m where severe winters are experienced. It is best to sow the banks with grass, but do not forget to keep it cut! Keep cattle out with a stock-proof fence and make all outlet points vermin-proof.

Break-pressure Tanks

Where water is being moved by pipeline over hilly areas the difference in head through the system may be so great that it is beyond the strength of the pipes. If this is the case, bursts will happen and there will be leakage from joints and valves. To control the pressure in the system a break-pressure tank is installed which brings the water into contact with the atmosphere again. As a result, the pressure in the pipes leading out of the tank is reduced. Figure 7.25 shows how a break-pressure tank works and the things to

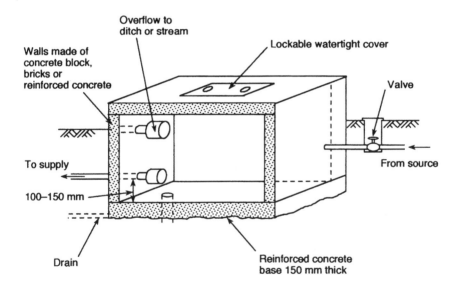

Figure 7.25 Break-pressure tanks are used to reduce the pressure gradient along a pipeline, thereby avoiding frequent bursts and leaks. They may also be used to control the delivery head in a hydraulic ram installation. A break-pressure tank is simply a tank which is connected to the atmosphere and allows part of the water flow to discharge to waste. The diagram shows the main features of such a tank with inlet and outlet pipes, a large-diameter overflow and a drain to facilitate cleaning. Make sure that the overflow pipe is large enough to take all the excess flow and ensure that the end is covered with a vermin-proof mesh. Where the inlet pressure is high you must have a heavy enough tank to stop it being blown off the end of the pipe. A solid concrete structure, partly buried below ground, will normally suffice. The features of a break-pressure tank may also be combined with those of a reservoir to provide extra storage within the system

consider when you are designing one. Another advantage in having a break-pressure tank in your distribution system is that the grade of pipe you use will not need to withstand high pressures and will be cheaper. High pressure pipe is usually much more expensive than standard pipe.

There is often no need to build a separate break-pressure tank in addition to your storage reservoir or tank. Provided you are able to fit an overflow which will allow the water from the input side to continuously run away, the break-pressure effect can be achieved.

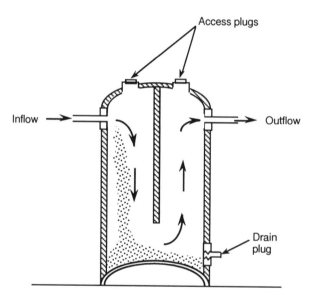

Figure 7.26 The ability of flowing water to carry particles in suspension depends on its velocity. Where water is flowing through a pipe its velocity at any point depends on the pipe diameter, with the highest velocities in the smallest pipes. This principle is used in settlement tanks to encourage material to fall out of suspension. An arrangement typical of domestic-sized tanks is shown in the diagram. Water enters the tank and is directed to flow round the tank to the outlet by a baffle plate which divides the tank in two parts. The larger particles will fall out quickly with the smaller ones needing the extra time provided by the slow velocities to settle before the water re-enters the pipe network. A drain plug is provided to enable the sediment to be drawn off. This must be done routinely on a frequent basis as the tank will not be effective once the depth of sediment reaches more than about halfway to the baffle plate. Additionally, any organic material in the sediment may decay, polluting the water. If all the material does not settle in the tank either replace it with a bigger one or use two in series

Sand Traps

A settlement tank (Figure 7.26) works by reducing the velocity of water flowing through it. The size of particles suspended in water depends upon the velocity. In other words, the faster it flows, the greater the size of particles it can carry. As water flows from a pipe into a tank the sudden large increase in cross-sectional area causes velocities to be reduced significantly but the flow rate remains constant. This may seem surprising, but if you mentally compare water flowing through the tank with that moving along a piece of pipe the same length, there is simply a large volume of water moving slowly through the tank and a small volume moving quickly in the pipe. The sudden drop in velocity causes the suspended material to settle out and fall to the bottom of the tank. Install the outlet pipe above the bottom of the tank so that it does not become buried by silt. Fit a strainer to the end of both the inlet and outlet pipes to reduce the silt getting into the system to a minimum. These strainers will also prevent unwanted vermin entering the tank.

DISINFECTION

Before you use any part of your water supply system it should be disinfected to prevent supplies from becoming contaminated with bacteria that may have entered the system during construction (Chapter 6).

8
Maintaining your System

When you have constructed your water supply system it is important to carry out regular maintenance so that it continues to function. If you have constructed your source works and distribution system properly, no serious problems should arise. No water supply system will last for ever but planned maintenance will extend its life significantly. Regular maintenance will also prevent water quality problems from developing which will keep your treatment costs to a minimum. You should undertake a regular inspection of your source and distribution system at least once a year and preferably twice a year. This chapter gives a few pointers towards the things which you should check at each routine survey. Most of the problems which can arise are easily put right if early action is taken. Delays may involve more extensive work and cost you money.

RAIN-WATER COLLECTION

It is important to clean out the silt-trap frequently so that the build-up of the silt/sand deposit does not become so large that the trap will no longer be effective. This would cause silt and/or sand to be washed into the storage tank. A more serious problem arises if the silt contains organic material which breaks down producing an unpleasant tasting material and harmful bacteria which may contaminate your water supply. To avoid this problem check the silt-trap every week and clean it out when necessary. Do not forget to disinfect the trap with a mild chlorine solution (see Chapter 6) after cleaning, taking great care not to contaminate the water in storage while you do it. Check your storage tank following the advice set out below.

SPRING COLLECTION CHAMBERS

Have a look at the stock-proof fencing and check that it is undamaged. Make any necessary repairs to the fence, posts or wire. Clean out the cut-off ditches and make sure that surface water can still get away without flowing towards the spring collection chamber. Have a look at the chamber itself and ensure that the structure is sound. If there is silt in the chamber bottom, clean it out but do not forget to close the valve on the outlet pipe first to avoid silt getting into your supply system. After you have cleaned out the chamber it is a good idea to disinfect it using a chlorine solution (see Chapter 6) before using it again. Check that the covers are still watertight and lockable and see that all your vermin-proof screens are still in good repair.

WELLS AND BOREHOLES

The most important thing to check here is that the lining material is still in good order. Examine the walls of your well for roots or stains which could indicate that pollutant material is getting into the well. Repair any cracks to the well walls with mortar and have a look round for the source of any pollutant. Remember that septic tanks, cess pits and sewage pipes are the most common cause of serious pollution in shallow wells. Make sure that the concrete apron surrounding your well or borehole is not cracked and that water will still run off away from the well. Clean any rubbish or debris away from the top of the well or borehole, and if your borehole is covered by a chamber, make sure that this is also clean. Have a look at the valves and fittings which are situated near the well head and make sure that they are still working.

If your well contains silt, this can be cleaned out in the same way as a spring collection chamber. Make sure that both you and your boots are clean and disinfected before you get into the well and that you take all the necessary safety precautions (see Appendix 2). If you are climbing into your well always have someone at the surface to help you if you get stuck. Check that the lining has not become loose before getting into the well. You can inspect the sides from the surface using a powerful torch or a shaft of sunlight reflected from a mirror. It is also very important to make sure that there are no harmful gases in the well. Old wells, especially if they are more than 15 m deep, can contain enough carbon dioxide to suffocate anyone who goes down them. One tried and tested method to prove the breathable quality of an atmosphere in a confined space is to lower a lighted candle into the well and observe the flame. If it flickers or goes out, the well must be ventilated before you go down. This can be achieved by lowering a large bundle of sacks, rags or other material down the well and quickly pulling it out again. This is repeated several times until the well is clear of gas. Repeat your candle test to make sure that the well is clear.

An alternative approach is to ventilate the well by pumping fresh air into it. You will need a fan coupled to flexible air ducting pipe. This pipe is about 150 mm in diameter and is kept open by a wire coil. Both the fan and ducting hose can be obtained from most equipment hire shops. Simply dangle the air pipe down the well and pump air into it for at least half an hour before you climb down to start work. Remember to check the atmosphere before venturing in.

In areas where methane may accumulate in the well it is vital that you do NOT use a light to test the air as you may end up blowing up both you and your well. Methane is commonly found in areas where coal is mined and can also result from the decomposition of organic materials like peat or refuse. In the case of peat the methane is termed marsh-gas or will-o'-the-wisp and can cause flickering lights on marshy areas as it burns. As refuse and other organic wastes rot down in a landfill they also produce methane. In this case it is termed landfill-gas. If your well is in an area where methane may accumulate, try to use a special methane detector to measure the gas in the well. If you cannot get hold of one, play it safe and ventilate the well before climbing into it.

Desilting of boreholes is a specialist job and usually requires the services of a drilling contractor. He will clean the borehole out, possibly by bailing it, using a similar technique to that described in Chapter 5. Boreholes can also be successfully cleaned out by "vacuum cleaning" the bottom using air-lift pumping techniques. An air-lift pump is set up as described in Chapter 3 with the intake set just above the level of the silt in the borehole. The pumping action will disturb the upper layers of silt and bring them to the surface with the pumped water. Once the water has cleared and been allowed to pump for five minutes or so, the pumping main should be lowered a little further into the borehole so that it is again 25–30 cm above the level of the silt. In this way all the silt is gradually removed from the borehole.

BOREHOLE EFFICIENCY

In Chapter 3 under the section on pumping tests, the concept of specific capacity of a well or borehole was described. This is the relationship between the drawdown and pumping rate. In many instances the walls of a borehole will slowly clog and this reduces the permeability of the well face causing a greater drawdown for each pumping rate. This clogging process can seriously reduce the yield of a borehole and also increase the cost of pumping water. If regular checks are made on pumping water level and rest water level on a basis of every six months or so, any change in specific capacity will be detected.

When water levels are measured it is important to make sure that the same period has elapsed since pumping either started or stopped, depending on whether the water level is a pumping level or rest water level. Ideally, a period of a day or two should be used but a rest water level cannot be obtained without shutting down the supply and it may not be worth this inconvenience. An adequate result can be obtained if a two-hour rest period is used each time. To compare the specific capacity, calculate the drawdown and plot this on the specific capacity graph that was drawn from the result of the test carried out at the time the borehole was constructed.

A reduction in the specific capacity value will indicate that the borehole should be cleaned out; this is a job for the drilling contractor, once again, who has several methods available to him. These include surging (as described in Chapter 5), high-pressure jetting where a rotating jet of water is used to scrub the borehole wall, and acidising to dissolve the minerals which have been deposited on the well face or screen. In some cases, where the borehole has no screen, the driller may use explosives to cause additional fracturing of the rock, thereby increasing the well yield. This is an exciting process to watch but it may not always work, and in any case the yield may often slowly deteriorate again, requiring a repeat performance in a few years time.

STORAGE RESERVOIRS AND TANKS

Check these structures for cracks and damage which could lead to leaks or allow the entry of contaminated liquid or vermin. Inspect the control valves, ball valves, level indicator, overflow, vents, manhole covers, and the general condition of the lining of the inside of the reservoir. Clean out any silt and again make sure that the valve on the delivery pipe has been closed to prevent silty water from getting into your supply system. Once all repairs and cleaning work have been completed, disinfect the reservoir with a strong chlorine solution. On the outside of the reservoir make sure that covers are still waterproof and the locks work; trim the banks and surrounding areas of grass and weeds; and repair any damage to the protective fences.

If you use a storage or break-pressure tank, check its condition in the same way. Make sure that the supports are not corroded and carry out any necessary repairs. If the tank or supports are rusty, clean and re-paint them but make sure that all paint which will be in contact with the water is non-toxic. If the tank is in an exposed position, make sure that the lagging and other frost-protection material around the tank and on all exposed pipes is adequate and in good repair. Do not forget that if the inside of the tank needs any attention it should be disinfected before being put back into use.

If you are unlucky enough to find a dead animal in your tank or reservoir, remove it and drain the tank. Refill the tank with a strong chlorine solution as described in Chapter 6 and shut off the inlet valve. You will also need to disinfect all pipework which may have been contaminated. To do this, open the valves and taps so that the disinfectant drains from the tank into the pipes. Allow the water to run until you have a strong chlorine smell at each outlet. Top up the tank with extra chlorine solution and leave it to stand for 24 hours. Finally, drain all the disinfectant from the system and refill with clean water. Allow this to drain away and repeat until the chlorine smell has gone. You may need to repeat this procedure in your house and at each building where the plumbing system has a header tank in the roof.

If it is necessary to enter a tank or reservoir to clean it, take the same safety precautions as for wells. Confined spaces are dangerous places, with possibly bad atmospheres and difficult escape routes (see Appendix 2).

PUMPS AND ASSOCIATED EQUIPMENT

If you have a pump house, make sure that it is weather-proof and vermin-proof, that it can be locked and is clean and tidy. Check that any electrical switchgear is working correctly; that it is still safely installed and wired; and that the control panel is dry and firmly secured to the wall. If in doubt call in a qualified electrician, but most importantly make sure that before carrying out any checks on electrical equipment you have switched off at the mains first. See that electric pumps are safely wired and earthed, and that all moving parts are protected by guards and safety cages. All lagging and other frost protection should be checked and renewed where necessary. If you use a petrol or diesel motor, service it regularly as recommended by the manufacturer because, as with your car, regular maintenance will prevent breakdowns.

Hydraulic rams should be inspected regularly to check for leaks around the joints. Listen to the way the pump is operating. If it sounds loud the air vessel probably needs topping up with air. At least once a year stop the ram to clean it. This will entail removing the air vessel and impulse valve. Check for signs of wear and replace worn parts. Make sure that you tighten the bolts fully when re-assembling and start it up again as described in Chapter 7.

Follow the manufacturer's advice on checks for your submersible or surface pump. Pump output should be checked occasionally using a water meter if you have one installed or by using the "jug and stop-watch" method, taking your measurement at the point where the water discharges into your storage tank. If there is even a very small sediment content in the pumped water, it may be sufficient to cause pump wear. The main symptom will be a reduction in the quantities pumped which can be confused with a loss of yield in wells and boreholes. Any other maintenance on the submersible pump will require the supplier to remove it and carry out the repairs. Most manufacturers of surface pumps recommend regular inspection without dismantling. Carry out any greasing or lubrication as recommended, keep the case clean, remove any rust and re-paint where necessary. Be careful not to spill the fuel or lubricant so that it could contaminate your water supply.

If the system includes a pressure tank, it is important to carry out regular maintenance because they can be a safety hazard. Carry out regular checks on the safety valves, replenishing or air valves, level indicators and gauges to make sure that they are working properly. Inspect your cylinder, looking for signs of corrosion and leaks. Coloured stains on the outside of the tank are a usual indication that slow leakage is taking place. In most cases it will be better to replace the tank rather than trying to patch it up.

PIPEWORK

The pipework distribution system should be included in your inspection programme. It is likely that most of the pipework is below ground but this does not prevent you from walking the route of the pipeline to look for any signs of leakage or areas where damage could have occurred. A likely place to look for leaks is at joints between lengths of pipe or fittings such as valves. If you have followed the advice in Chapter 7 and built a small chamber over the joints to keep them accessible you will be easily able to check for leaks. Valves may leak if the washer has become worn or the valve spindle needs re-packing. A good plumbing guide will help you carry out these simple tasks.

When you walk along the pipeline route look out for signs of a leak. Water pouring out of the ground is an obvious sign of leakage, but lush green vegetation may also be a give-away. Water flowing in a drain at times when it is not expected may also be a useful sign. If you suspect that a section of pipework is leaking, you can carry out your own checks. One way is to install a water meter at each side of the suspect section. Take regular readings over a period of a few days and compare the quantity of water which has entered the length with that which has flowed out through the meter at the other end.

The above method may not be a very convenient way for the average person to carry out these checks but it is possible to listen for water leaks using a traditional water industry technique. All water leaks cause a small noise which can be heard even if it is buried a metre down. The way to find a leak in this way is to use a length of stick such as an old broom handle which is placed against the ground. Place your ear against the other end and listen carefully for a low hissing sound made by the water squirting out of the pipe into the ground. Unless you live in a very remote and quiet area the best time to carry out this sort of survey is late at night when everyone else is in bed. It is surprising how accurately leaks can be pin-pointed in this way and, although more sophisticated electronic gadgets are available, these techniques are still used by water professionals.

Once the leaking section of pipe has been identified, dig down, cut it out and replace it with a good piece. If a section of pipework consistently produces bursts or leaks it would probably be better to replace the whole section rather than continue repairing it. Replacing problem lengths of pipe is a good idea because when leaks start to happen it often indicates that the entire pipe system has become weak. In this situation once you have repaired one leak another will start up as the next weakest piece of pipe fails and you may as well replace the whole pipe run. Make sure that any new pipe is of a class and standard suitable for the working pressures of the system as this is probably the reason you are having the problems. It is also a good idea to take the opportunity to replace metal pipe with plastic ones which have a number of advantages, as explained in Chapter 7.

Check all exposed pipework for leaks and frost protection. If your system includes outside taps and stand-pipes make sure they are securely fixed, look for leaks and see that taps are working properly. Have a look at the condition of any valves and stop-cocks and make sure that they still work. Make any repairs to valve chambers, including lids.

Do not forget to include the indoor plumbing system in your routine inspection. Replace the washers on leaking taps as soon as they start. The steady drip may not seem much but remember that you have paid with your hard-earned money to pump and treat the water only to let it run to waste. Estimates of such losses show that 31 500 litres will be wasted each year with a leakage rate of only 1 ml/s. This represents a fast drip, but even a slower one of 10 ml/min will lose over 5000 litres in a year.

TREATMENT SYSTEMS

If you use a treatment system to disinfect or soften your water supply, regular routine maintenance is essential. Lack of maintenance will mean that you will be drinking untreated water without realising it and health problems most likely involving diarrhoea are likely.

Manufacturers of water treatment systems will provide information on the routine maintenance checks which are required for each piece of equipment. Filters must be back-washed regularly and softening plants should be reactivated at the prescribed intervals. It is easy to let the regular replacement of filter cartridges slide but do not be tempted to ignore this chore. Bacteria commonly build up in the filtrate and can cause greater problems than if you had not bothered to filter the water in the first place.

9
Water Rights

The legal right to abstract water has been of fundamental importance to all people since time immemorial. The system varies from country to country although generally land ownership or occupation gives a right to abstract water from streams or wells on the property. Because of the diversity of the legal system round the world I am unable to give details of water rights for all countries. The systems which apply in the UK and the USA are covered in some detail to give you an idea of the sort of arrangements which apply. I have also made a few general comments on the laws and regulations in several other countries so that you can see the sort of legal requirements which may apply where you live.

Before we get into the complexities of water rights, we ought to consider the law in general. The first thing to remember is that the law is very complicated. Much depends on the interpretation of words and sometimes the use of otherwise archaic terms. This means that this chapter is going to contain some legal terms and phrases and I apologise for the rather strained style. If in doubt, please refer to the glossary where the most important terms are listed. For example, groundwater is often referred to in legal circles as "underground water" or "water contained in underground strata".

OBTAINING WATER FROM OTHER PEOPLE'S LAND

The legal system of water rights described in this chapter is based on the occupier of a property obtaining water on his own land. This is not always the case and many small properties obtain their water from sources on other people's land and quite often the right to this water is covered by the deeds of the property. Should you be thinking of buying a country cottage or smallholding which does not have a water source on its own land, you ought to check the deeds very carefully before buying.

There are many different arrangements for obtaining water supplies from other people's land. You may have a straightforward right to the sole use of a spring. Alternatively, you could be one of a number of people obtaining water from a spring or reservoir. Sometimes neighbours may share the cost of building a source, such as a new borehole. They may be on such good terms that no formal arrangements are made, which works very well until they fall out or one moves and new people move in next door. It is essential that the rights to a supply of water are set out clearly in a written agreement which will form part of the deeds.

ESTATE SUPPLIES

Sometimes groups of country properties are supplied with piped water by an estate company acting rather like a water board. In the USA there are thought to be 140 000 such small systems which serve less than 25 people or have less than 15 piped connections to different properties. Often good quality, reliable water supplies are provided by these small schemes at a reasonable cost.

There are instances, however, when lack of proper maintenance results in the estate supply being unreliable or even polluted and unsafe to drink. The local Environmental Health Department will put pressure on the estate company but the arguments may drag on for months or even years, while you are suffering the bad supply.

One solution to all problems associated with water supplies from other people's land is to have a borehole drilled in your back garden. Remember that in many places around the world the odds are in your favour to obtain enough water to meet your domestic needs from a well or borehole. The main disadvantage is that you will have to cover the costs yourself but it may be worth it for peace of mind.

THE UNITED KINGDOM

There are two main divisions in the British legal system; these are *common law* and *statute law*. In some ways statute law is more easily understood as it is based on the laws passed by Parliament. Common law, on the other hand, is not set down in Acts of Parliament. It is the law of precedent and has evolved over the years from the decisions of judges made in courts.

The rulings made by judges in court actions are used to resolve new cases with similar circumstances. If you read accounts of court cases in the newspapers, or watch courtroom drama on the television, you will remember barristers quoting the cases of *Jones* v. *Smith* or *The Crown* v. *Bloggs*. These are references to decisions made by a judge in earlier cases and much of the argument surrounding a present case will be to decide which previous rulings are now relevant. The two parts of the legal system are quite separate. Statute law takes precedence over common law.

Common Law Water Rights

In the British Isles the fundamental rights of people or companies to abstract water have been built up over the years, mainly based on English court cases. Common law rights are the same in England, Scotland, Wales, Northern Ireland and very similar in the USA, the Irish Republic and many other countries. However, since 1963 these rights have largely been superseded by statute law in England and Wales.

Riparian Rights

Persons who occupy land, either as the owner or tenant, have what are termed *riparian rights*, which is the right to the benefits of water flowing in a stream across the land or along its boundary. The riparian doctrine gives a right to share water with the other riparian owners. It does not give a right to a specific volume or flow of water. In general, each riparian owner may use as much water as they need for any reasonable purpose. These benefits mean the use of water in its natural state, which applies to both the quantity and quality of the water. Riparian rights go with the land, and are enjoyed by anyone who legally occupies the land and are not forfeit if the water is not used for a period. In the UK the riparian use of water is divided into two general classes; these are *ordinary use* and *extraordinary use*.

Ordinary Use

Ordinary use is the general case, where the occupier has the right to the reasonable use of water flowing past his land to supply his house and to water his cattle and other livestock. There is no restriction as to how much he can take. He can use all the water in the stream. Downstream riparian occupiers are not entitled to complain even if they are deprived of water in this way.

Extraordinary Use

Extraordinary use is when the water is used for industrial purposes. It also allows a stream or river to be dammed or diverted to drive a water mill and to use the water for irrigation (but not spray irrigation). This time, however, a riparian occupier must not interfere with the rights of riparian occupiers both upstream and downstream of the land in question. The water must be used in connection with the riparian lands and must be returned to the watercourse substantially unchanged in both volume and quality. Common law rights do not recognise the abstraction of water for spray irrigation because water is not returned to a stream in an unaltered condition. An additional restriction is that the use to which the water has been put must be "reasonable" — a concept that could keep lawyers arguing for days.

Effects of Land Drainage

There are two special cases which relate to land drainage as it affects the use of water. Once the owner of land on which a spring issues has had 20 years uninterrupted use of this spring, he (or she) has established an absolute right to that water. Should his neighbours dig ditches or trenches which reduce the flow of water from his spring, he would have a right of action against them.

The situation is somewhat different for a person who uses water from an agricultural land drain on his land. Any rights he has to use this water do not prevent the owner of the drained land from improving the drainage system, even if this cuts off the water supply.

Underground Water

The common law rights to use groundwater are the same as for surface water when it can be shown that the groundwater is flowing in defined underground channels. This only applies in a few cases. Water running through abandoned mine workings or well-established cave systems would come into this category.

Where water is percolating through the ground, no riparian rights exist. The occupier of a piece of land can sink a well or drill a borehole and, if his abstraction dries out his neighbour's well or spring, the neighbour has no right of action against him.

Loss of Support

It has been established in common law that the owner of a piece of land cannot withdraw support from his neighbour's by removal of soil. If you decide to dig an enormous hole and your neighbour's house falls into it, he will be entitled to look to you to make good the damage. No one, however, has any right of support from water which is contained in his land. Should the consequences of pumping groundwater from a well or borehole cause subsidence, there are no rights of action. On the other hand, should the water which is pumped from this well contain sand or silt in any quantity, things are different. This constitutes loss of support and the person pumping the water would be liable to pay damages.

In England and Wales, these common law rights to abstract are no longer as important as they once were as they have been superseded for major abstractions by the Water Resources Act 1991. Common law still applies when abstraction licences are not required.

STATUTE LAW

Prior to 1963 the only Acts of Parliament which applied to water abstraction were private Acts. These authorised named water boards, water companies or industrial companies to abstract water for specific purposes. Since the mid 1960s there have been two Acts of Parliament which seek to control water abstraction to some extent.

In Scotland, river purification authorities can control the abstraction of water from streams for spray irrigation under the provisions of the Spray Irrigation (Scotland) Act 1963. The Act enables river purification boards to apply to the Secretary of State for Scotland to limit water abstractions from specified streams. Once an order has been made, it is necessary to obtain a licence from the river purification board to abstract water for spray irrigation. Licences are granted for a year at a time and there are special provisions to cover a lack of water and floods. This is the only private water abstraction that is controlled in Scotland by statute law. All abstractions for purposes other than spray irrigation are subject to common law rights.

There are no statutes which control abstractions in either Northern Ireland or the Republic of Ireland.

Water Resources Act 1991

The abstraction of water in England and Wales is controlled by the Water Resources Act 1991, which superseded similar legislation contained in the Water Resources Act 1963. In simple terms, the Water Resources Act requires all abstractions of water to be controlled by a licensing system administered by the National Rivers Authority (NRA) although there are a number of exceptions where a licence is not needed.

These exemptions are important to the private abstracter because small abstractions from surface waters which run through or border your land do not need a licence provided that you take less than $20\,m^3$ in a day for domestic or agricultural use. Abstractions from wells and boreholes for the domestic use of one household are also exempt from control for the same quantities. A licence is needed, however, where the water is used for agriculture. Other exemptions exist for one-off abstractions up to 5 m³, or 20 m³ with the NRA's consent. No licence is needed for test pumping a new well provided that the NRA has given a consent and water for fire fighting is also exempt.

Licences are required for spray irrigation, industry, trade and most commercial uses. Impoundings for whatever purpose are also likely to need a licence. If in doubt, check with the local NRA office. Remember that by not having a licence when you need one, you lose the protection of the Water Resources Act as well as risking prosecution.

Protected Rights

The main innovation of the Water Resources Act is that it establishes what are known as Protected Rights. These Protected Rights basically mean that no one can come along and start to make a new abstraction which interferes with any existing ones. The new abstraction requires a licence, however; otherwise the common law principles apply.

Protected Rights cover both licensed abstractions and those which are exempt from licensing. The Water Resources Act makes the safeguarding of these Protected Rights a duty of the NRA. The NRA is forbidden to grant a new licence which will stop any existing protected abstractions from being made. This interference between sources is called *derogation*. If the NRA should fail in this duty, the injured party can take action against the NRA instead of the abstracter who is causing the problem. As you can see, this contrasts with the common law situation where any rights of action are between individuals.

Who can Apply for a Licence?

If you want to apply for a licence to abstract from a surface water source, you must be the riparian occupier of the adjacent land. Anyone who is about to buy or rent such land can also apply for a licence. In some circumstances a right of access to the source may be sufficient.

If you buy or rent some land for which there is an existing licence to abstract water, you can apply to have the licence transferred to you. It is important to contact the NRA as soon as possible after you buy the land, ideally before you sign the papers as it is likely that you will only have one month to transfer the licence into your name. If you miss the deadline you will have to go through a full licence application and it will not be certain that you will be granted the licence as you will be treated as a new applicant.

To apply for a licence to abstract from a well, borehole or other excavation which contains groundwater, you must be the occupier of the land *comprising the underground strata* from which you want to abstract. This means that the well, the borehole, etc., must be on your land.

Application Procedure

The application for an abstraction licence comprises several elements. You must complete an application form which is fairly straightforward as far as official forms go. It comes in two parts and the NRA have put a lot of effort into making it simple to follow. You must also provide a map to show the position of the proposed abstraction and the land you occupy. This map is usually on a 1:10 000 scale and you will be asked to provide two copies. You will also have to pay an Application Charge to defray the cost of the NRA processing your application.

In some cases the NRA will require that you have an environmental assessment carried out to examine the environmental impact that your abstraction is likely to have. In straightforward cases you will be able to provide the information that the NRA require; otherwise you will need to employ a consultant who will do it all for you. Make sure that the person you employ has experience of completing this type of assessment before you give them the job.

The most important part of the application is to advertise in a local paper and the *London Gazette* (the official government newspaper) that you want to take water. The procedure laid down by the Act requires the advertisement in the *London Gazette* to be within three days of it being published in the local paper. The advertisement in the local paper must be repeated in the following week but this is not necessary for the one in the *London Gazette*. The notice in the paper must give full details about the proposed abstraction and also specify an address in the locality of the abstraction where a copy of the application and plan can be inspected by any member of the public. The application has to be available for a period of at least 28 days from the date when the notice first appeared in the local newspaper. The details must include your full name and address. You must name the stream from which you are to take the water or must specify where the water is to be abstracted; for example, state that the abstraction is to be from a well, borehole or spring, etc. The national grid reference of the abstraction point or points must be given, as well as the address of the property. The NRA provide a helpful form setting out these details for you to complete and send to the newspapers.

The maximum quantities you want to abstract on an hourly, daily and yearly basis must always be quoted in the advertisement. The precise form of the advertisement is described in regulations made under the Water Resources Act and it is essential that you get it right, otherwise you will have to go through the procedure all over again. This will cause delay in your application and a lot of extra expense as your advertisements are likely to cost you well over £200. Fortunately, help is at hand. When you are employing your own expert you can expect him/her to get it right; if you are doing it on your own, however, you will find that your local NRA will give you advice and help you to specify things properly. It is a good idea to let them see a copy of the advert before you send it off to the papers. An example of an advertisement is given in Figure 9.1.

Once you have placed your advert and the appropriate periods have elapsed, you send your completed form to the NRA, together with copies of the map and the advertisements which appeared in the local paper. These copies must be actual copies of the newspaper but there is no need to send a copy of the *London Gazette* as the NRA gets its own copy.

All you have to do now is sit back and wait for the NRA to make up its mind. This may not be too long as the Act requires a decision to be given within three months after the date the authority received your application. This might seem rather a long time, but the licensing staff must thoroughly investigate your application before they can make a decision. They have to make a good job of it to avoid granting a licence which would derogate from someone else's Protected Rights.

The NRA must take into account any objections which they have received in writing within the prescribed time after your adverts

---

WATER RESOURCES ACT, 1991

NOTICE OF APPLICATION FOR A
LICENCE TO ABSTRACT WATER

Take notice that John Edward Smith of Plumtree Farm, Greasley, Newtown is applying to the National Rivers Authority for a licence to abstract water from Triassic Sandstone by means of a borehole, at National Grid Reference JG 346573, Plumtree Farm, Greasley, Newton.

The proposal is to abstract water at the following rates:

2 cubic meters per hour; 5 cubic metres per day;
and 1800 cubic metres per year.

The water will be used for domestic and agricultural purposes.

A copy of the application or of any map, plan or other document submitted with it may be inspected free of charge at all reasonable hours at Plumtree Farm, Greasley, between 24 August, 1995, and 21 September, 1995.

Any person who wishes to make representations about the application should do so in writing to the National Rivers Authority, North Central Area Office, Edwards House, High Street, Newtown, before the end of the said period.

Signed: J. E. Smith
19th August 1995

Figure 9.1   It is important to use the correct form of words and include all the relevant information in an advertisement for a new abstraction licence. This example will give you an idea of the format but make sure that you check it with the local NRA licensing staff before you place your advert in the paper

appear. Their investigations may take some time to complete, especially if field measurements and tests are required. Sometimes they cannot come to a decision within the three-month period and in that case they will ask the applicant to agree to an extension. Perhaps it may be necessary for information to be collected on dry weather flows in the stream you want to use for your supply, or perhaps the pumping test on your new borehole needs to be carried out during the summer, which will cause a delay.

If you feel exasperated about the amount of time and effort that the NRA is devoting to your application, remember they are looking after the interests of the existing users and when you get your licence you will be given the same degree of protection.

Besides ensuring that no existing abstracter suffers from derogation, the NRA must take into account several other things. They will
decide whether or not the quantities you want to abstract are
reasonable to meet your various requirements. If you have assessed
your water needs too high they may grant you a licence to take
reduced quantities. Sometimes things may go the other way and the
NRA may suggest that you need greater quantities than you realise.
This must be done at the pre-application discussion stage or you will
have to re-advertise all over again. This is yet another reason to
contact the water authority at an early stage.

Environmental Impact

Part of the deliberations of the NRA on all abstraction licence
applications is to assess the size of any impact which the new
abstraction may have on the local water environment. If they think
that the impact may be large they will ask you to carry out an
environmental impact assessment. In most cases you will need to
employ a consultant to do this for you. The NRA staff will give you
the names of a few local consultants who are able to help. Do not
forget to check them out to see if they have done this sort of thing
before and if their old clients are happy with them. An example
where such an impact assessment would be required is where a
pumping from a new well is thought to divert groundwater feeding
into a natural boggy area, thereby altering the local habitat for
plants and animals.

Minimum Acceptable Flow

Where someone proposes to abstract from a stream or river, the
NRA must take into account the *minimum acceptable flow*. This
concept takes into account that a watercourse is used for many
things. Besides being a source of water for common law and
licensed abstracters, it may also be used for fishing and navigation.
From a public health point of view it may be necessary to ensure
that there is always sufficient water in the stream to dilute sewage
discharges. When fixing a minimum acceptable flow, the NRA will
take these considerations into account. They must also take account

of any natural beauty of the landscape where the stream or river is an important factor.

Licence Provisions

When you eventually receive your licence, you will find that it sets out what you are permitted to do in some detail. The maximum quantities which you can abstract will be stated for any hour, any day and the yearly total. Should it be thought necessary to restrict abstraction to specific periods during the year, these will also be stated on the licence. These restricted periods are quite common when water is used for spray irrigation or frost protection. You will be expected to measure and record the quantities you take and the method to be used may be specified on the licence; if not, you will have to agree the method with the NRA.

The means by which you will take the water will also be specified. Usually, this will be a pump of some sort and the capacity of the pump will be stated. This means that you cannot change your pump for a bigger one without applying to vary your abstraction licence. The licence will also specify the name of the licence holder, the land where the water can be used and any expiry date. If a date has not been defined, it will be stated on the licence that it remains valid until revoked. This is the usual case and means that the licence will last as long as you want it. If you sell your land, the new owner can succeed to the licence but he must notify the NRA within one month of taking occupation, otherwise the licence will automatically lapse.

Altering a Licence

If you want to change any of the conditions stated on your licence, you will have to make an application to the NRA following the same procedure as when you applied for the licence in the first place. This includes advertisements and waiting the prescribed periods to give objectors the chance to write to the NRA. There is one exception to this requirement and that is where you want to reduce the quantities on your licence. In this case, all you need to do is to write to the NRA stating the reduced quantities and your reasons for reducing them.

Right of Appeal

You have the right of appeal to the Secretary of State for the Environment if your application is refused by the NRA or if a licence is granted subject to conditions which you do not like. If the NRA asks for an extension of time to consider your application and you do not want to grant it, you can also appeal to the Secretary of State. He has the final say in all matters relating to water abstraction licences.

Your case will be considered in great detail by the engineers in the Department's Water Directorate and they will advise the Secretary of State. These people are very thorough and look at every aspect of each case in great detail. You will be asked to set out your case, as will the NRA, and you will each get the chance to comment on each other's evidence. After that, each side is allowed to respond to the comments the other has made.

The civil servants involved may decide to seek expert advice, usually from another government department, so you will not be surprised to learn that it can take a long time to resolve such an appeal. In some instances the matter would be investigated at a public inquiry but this only happens when there is a lot of opposition or when one side demands it. In most instances an appeal is settled by exchange of evidence in writing.

You should never embark on an appeal to the Secretary of State without giving it a great deal of consideration and seeking professional advice. You will need legal advice as well as employing the appropriate technical experts.

Register of Abstractions

The NRA must keep a register of abstraction licences, applications, revocations, etc., at each office for the area covered from that centre. This register, which includes a set of maps showing the location of all the abstraction points, is available for inspection by the public during office hours. By examining the register you can get a good idea of how many abstractions there are in your area, but remember it does not include details of common law abstractions which may need an on-the-ground search. The NRA may have some information on common law abstracters for the area but it is likely to be incomplete. The local authority should also have information on private sources in their area and may be able to help you.

How a Licence Protects You

When you become a licensed abstracter, you will not be liable for causing the loss of someone else's supply. This applies where the abstraction is made as a Protected Right and also where you would have no defence under common law. This liability is passed on to the water authority, so you can see why they are so careful before granting a licence.

The NRA's liability only applies to people who hold Protected Rights and are affected by a licensed abstracter. This abstraction must be made following the conditions attached to the licence. In this part of the law, where either common law or statute law may apply in different circumstances, things need to be thought about quite carefully to avoid confusion. If you are abstracting water within the terms of your licence you will not be liable in "nuisance" to another riparian occupier even though you may have no defence in common law. On the other hand, if you contravene the conditions of your licence or abstract without a licence when one is needed, you may have a defence against depriving someone of his water under common law, but you will be liable for a statutory penalty for your infringement of the Water Resources Act.

If you are abstracting water without a licence because you are exempt, then your defence against an accusation of affecting someone else's abstraction will rely entirely on common law. The fact that your abstraction falls into one of the exemption categories only removes a liability for statutory penalties and does not provide a defence under common law. There is nothing in law to prevent you from applying for a licence to gain this protection but you would then need to comply with the provisions of the licence and pay any fees which may be due.

Licence Charges

The NRA charges an annual fee for abstraction licences based on the maximum permitted quantities in any year. Details of the charging scheme can be obtained from your local NRA office.

The charges are related to the maximum permitted quantities you can abstract so you may be paying much more than you need if these quantities are unrealistic. In special circumstances it may be possible to enter into an agreement with the NRA to reduce these charges. These agreements usually only apply where water is used for spray irrigation and allow the charges to be based partly on the licensed quantity and partly on the actual quantities used for a period of at least five years.

The law requires you to pay your charges within 14 days of a written demand being served on you by the NRA, who have the power to suspend your licence if you do not pay up. As these water charges are only about one-tenth of what you would pay for a piped supply and the licence gives you such splendid rights which protect your abstraction, you would be foolish to delay payment and risk losing your licence.

The money raised from abstraction charges is used to fund all the costs incurred by each region of the NRA in carrying out its water resources functions. The total amount collected only goes to meet these expenses as the costs of other NRA activities are paid from other sources of income.

Impounding Licences

The Water Resources Act 1991 requires all impounding works to be licensed, as well as any water abstraction. There are some exemptions where the impounding is controlled by another Act of Parliament or is looked after by a navigation or harbour authority. Should you want to build a little dam on your stream for any purpose, you need to apply for a licence to build it. When you wish to abstract from the reservoir the NRA can grant a combined licence which covers both the impounding and the abstraction.

There is no exemption if your dam is small; the wording of the Act is that no one can "obstruct or impede the flow of an inland water" and this covers all cases except those specifically exempt. Taken to its logical conclusion, this requirement for a licence even extends as far as stepping stones.

The procedures you have to follow when applying for an impounding licence are broadly the same as for an abstraction licence. The NRA will need to know what arrangements you are going to make for an overflow for both normal and storm conditions and they will also take into account the other land drainage effects and interests of riparian occupiers. You will also have to provide the NRA with details of the design of your dam. This must define the area you will be flooding and the volume of water which will be held back by the dam. If this volume is more than five million gallons then you will be subject to the provisions of the Reservoirs (Safety Provisions) Act 1930. This will mean that things are suddenly going to cost you a great deal more.

The Reservoirs (Safety Provisions) Act 1930 stipulates that only one of a select group of engineers can design and supervise the construction of the dam. Once it is built, you will have to have the dam inspected by one of these engineers every 10 years, and carry out any works that he (or she) may recommend to ensure its safety. It is a lot easier and cheaper to make sure that you do not have a reservoir that large.

The NRA is not responsible for enforcement of this Act. The County Councils are responsible for maintaining a register of large reservoirs in their area and ensuring that all provisions of the Act are met. The Department of the Environment's Water Directorate is responsible for overseeing this activity. The NRA engineer who processes your application will check the size of the volume you intend to impound and put you in touch with the local authority if it is over five million gallons. If you need to employ a suitably qualified engineer you will be able to obtain a list of approved engineers by writing to the Secretary of the Institution of Civil Engineers in London or the Department of the Environment's Water Directorate.

Groundwater Abstractions

The Water Resources Act 1991 makes it illegal for you to construct a well or borehole or extend an existing one without prior consent from the NRA. It is quite usual to start the long road to obtaining a groundwater abstraction licence by applying for consent to construct and test pump your new well or borehole. You should contact the NRA even if you will be using the water for your household and will not need an abstraction licence.

As with applying for an abstraction licence, there is an application fee, but in the case of consent to drill and test pump a new borehole there is no application form. The NRA will need to have details of the location, preferably on a large-scale plan. They will want to know to what depth you want to construct the borehole and to have an indication of the quantities of water that you would like to abstract from the borehole once it is constructed and licensed.

If you are given permission to construct and test pump your well or borehole, it will be granted by the NRA under Section 32 of the Act. This allows them to impose any conditions whatsoever on the consent. If you do not like the conditions, I am afraid that this is unfortunate as there is no right of appeal. Your only course of action would be to apply for an abstraction licence and when this is eventually refused you would then have the right of appeal to the Secretary of State for the Environment. The NRA, however, rarely imposes unreasonable conditions and only refuses to grant consent in certain rare circumstances. They may already know that abstractions from the aquifer you want to tap are already fully utilised, for example. In these circumstances a further abstraction could not be permitted and, as the NRA would not be able to grant a new licence, they would probably refuse to grant consent. In an appeal to the Secretary of State in these circumstances, you are likely to lose as he would support the NRA if their case is technically sound.

Consent Conditions

In granting a consent to you, the NRA will impose certain standard conditions. You will be required to carry out a survey to locate all protected sources within a 1 or 2 km radius from your borehole site. The maximum depth and diameter and, in some circumstances, the details of borehole casing or well lining will be specified. A consent is likely to set out the rates at which you can carry out your test pumping and require you to take a variety of water level and spring flow measurements to ensure that the effects of pumping from your new source are understood. The NRA will check to see that you are carrying out this work satisfactorily but it will remain your responsibility to ensure that you comply with the conditions of the consent and all the work required by the consent will have to be at your expense. There is no need to panic. In the majority of cases involving small abstractions the consent conditions are straightforward and it is not difficult to carry them out.

The order in which you apply for a consent and an abstraction licence is entirely up to you. It may be an idea to start the licence application procedure at an early stage so that the people in your area who have Protected Rights can make representations and have their sources included in the monitoring programme. This will also shorten the period between your initial approach to the NRA and being granted a licence. It is not really possible to generalise as to how this should be done and the local NRA staff or your consultant or contractor will advise you regarding the best order to carry these things out. The important thing to bear in mind is that you should start talking to the NRA about your ideas as soon as possible.

The purpose of the consent procedure is to allow the NRA to satisfy themselves that derogation will be avoided. In some cases, a well or borehole can be affected by having the water level lowered but derogation will not be held to have occurred if "permissible alterations" can be carried out by the plaintiff which would enable him to abstract the same quantities as before. The term "permissible alterations" means things like lowering the pumps or even perhaps deepening the well or borehole in some circumstances. If your source is affected in this way you are likely to end up out of pocket and you would be advised to see your solicitor at an early stage. This is particularly important if your abstraction is exempt from licensing, when the meaning of "permissible alterations" is not very clear.

The legal aspects of water rights can be quite complicated as you can see. In the majority of cases, however, people continue to abstract their water or apply for new licences and construct their wells and boreholes and stream intakes without having any problems. The chances are, therefore, that despite the legal maze you may have to walk down you will end up with a licence. If things do not go smoothly, you would be advised to sit back and think things through carefully before you take the next step. Water rights seem to raise very strong passions and I have known families and neighbours fall out over problems which could have been avoided if given due consideration at an early stage.

UNITED STATES OF AMERICA

There are many similarities between the system of water rights in the USA and those in the UK as they are both tied to the ownership of land. There are two different principles which control abstraction from surface streams in the United States; the *riparian doctrine*, which applies in the wetter East; and the *appropriation doctrine* which applies in the arid west. The right to abstract groundwater is more varied from state to state and includes the *rule of absolute ownership*, the *rule of reasonable sharing*, the *rule of correlative rights*, and the *appropriation* or *permit system*. The maps shown in Figures 9.2, 9.3 and 9.4 show which rules apply in each state.

Riparian Doctrine

Also often called the *rule of reasonable sharing*, the riparian doctrine is basically the same system as I have already described for the UK although there are substantial differences in the way the principles are applied from one state to another. In essence though, it is based

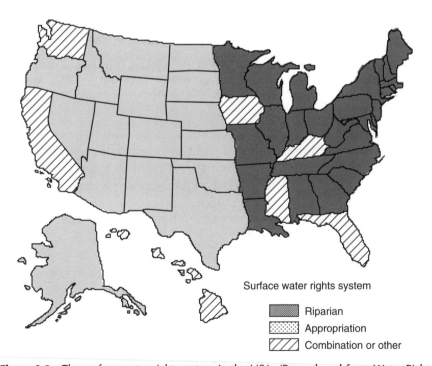

Figure 9.2 The surface water rights system in the USA. (Reproduced from *Water Rights of the Fifty States and Territories,* by permission. Copyright © 1990, American Water Works Association)

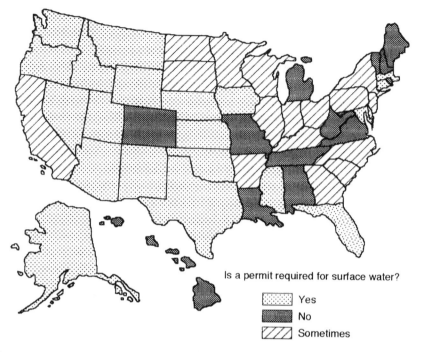

Figure 9.3   States where surface water permits are required. (Reproduced from *Water Rights of the Fifty States and Territories*, by permission. Copyright © 1990, American Water Works Association)

on the ownership of land which abuts surface water streams, lakes or other water bodies and gives a right to share water with the other riparian owners. Again, it does not give a right to a specific volume or flow of water and each riparian owner may use as much water as they need for any reasonable purpose.

There are limitations as to the place where the water may be used and the quantities involved. These vary from state to state and it will be important for you to find out which rules apply in your home state. In some states the place where water may be used is limited to the catchment boundary or may be fixed by the boundaries of the nearest quarter section in the original US survey, or by the historical boundaries of the riparian land which abuts a particular body of water. Not very simple, is it? Quantities are not defined in terms of specific volumes but are limited to being for reasonable purposes and require each riparian owner to leave enough for the others. No wonder there are arguments from time to time!

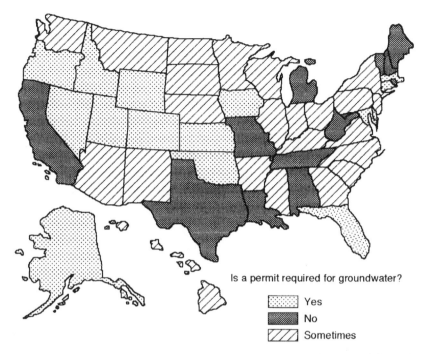

Figure 9.4 States where groundwater permits are required. (Reproduced from *Water Rights of the Fifty States and Territories*, by permission. Copyright © 1990, American Water Works Association)

Appropriation Doctrine

The riparian system became accepted in the East because the climate and availability of water allowed it. As settlers moved further west, however, they found very different conditions. It is significant that the early explorers called the Great Plains the "Great American Desert" and not everyone thought that it could be settled. Water was limited and as new settlers arrived conflicts arose over its use and availability. It is thought that the miners found the solution to the problem. They accepted the custom that the first to arrive and use water to work his claim would be protected against late-comers. Soon this principle became applied to all forms of water use in addition to mining. The custom became law by its recognition in court decisions over the years and was eventually enshrined in various statutes passed in individual states and by the Federal Government.

Table 9.1 Summary of water rights of the fifty States and Territories. Reproduced from *Water Rights of the Fifty States and Territories,* by permission. Copyright © 1990, American Water Works Association

| State or Territory | Surface water right system* | Permit required | | Court approval | | Buying/ selling | Disputes† | Number of compacts | Number of treaties | Legislative activity | Administering agency |
		Surface water	Groundwater	Surface water	Groundwater						
Alabama	R	No	No	No	No	No	GC	0	0	No	None
Alaska	A	Yes	Yes	No	No	Yes	WA, GC	0	0	No	Dept. of Natural Resources
Arizona	A	Yes	Yes	No	No	Yes	WA, GC	2	1	Yes	Dept. of Water Resources
Arkansas	R	No	No	No	No	Yes	GC	0	0	No	None
California	O	Yes	No	No	No	Yes	GC	3	1	Yes	Water Resources Control Board
Colorado	A	No	Yes	No	No	Yes	SC	11	1	Yes	Div. of Water Resources
Connecticut	R	Yes	Yes	No	No	No	GC	2	0	Yes	Dept. of Environmental Protection
Delaware	R	Yes	Yes	No	No	No	WA, GC	4	0	No	Dept. of Natural Resources
Florida	O	Yes	Yes	No	No	No	WA	0	0	No	Regional Water Management Districts
Georgia	R	Yes	Yes	No	No	Yes	GC	0	0	No	Dept. of Natural Resources
Hawaii	O	No	Yes	No	No	No	WA, SC	0	0	No	State Water Commission
Idaho	A	Yes	Yes	No	No	Yes	GC	2	0	Yes	Dept. of Natural Resources
Illinois	R	No	No	No	No	No	GC	0	1	No	Div. of Water Resources
Indiana	R	No	No	Yes	No	Yes	WA, GC	0	1	Yes	Dept. of Natural Resources
Iowa	O	Yes	Yes	No	No	No	WA, GC	0	0	No	Dept. of Natural Resources
Kansas	A	Yes	Yes	No	No	Yes	GC	4	0	Yes	Div. of Water Resources
Kentucky	R	Yes	Yes	No	No	No	WA, GC	2	0	No	Div. of Water
Louisiana	R	No	No	No	No	Yes	GC	2	0	No	None
Maine	R	No	No	No	No	Yes	SC	0	1	No	None
Maryland	R	Yes	Yes	No	No	No	WA, GC	3	0	No	Water Resources Administration
Massachusetts	R	Yes	Yes	No	No	Yes	GC	0	0	Yes	Div. of Water Supply
Michigan	R	No	No	No	No	No	GC	1	1	No	Dept. of Natural Resources
Minnesota	R	Yes	Yes	No	No	Yes	WA, GC	1	1	No	Dept. of Natural Resources
Mississippi	O	Yes	Yes	No	No	No	WA, GC	0	0	No	Dept. of Natural Resources
Missouri	R	No	No	No	No	Yes	GC	0	0	No	None
Montana	A	Yes	Yes	No	No	Yes	WA	1	3	Yes	Water Resources Div.

State	Type*					Courts†				Agency
Nebraska	A	Yes	No	No	Yes	WA, SC	5	0	Yes	Dept. of Water Resources
Nevada	A	Yes	Yes	No	Yes	WA, GC	1	0	Yes	Div. of Water Resources
New Hampshire	R	No	No	No	Yes	GC	0	0	No	Water Resources Div.
New Jersey	R	Yes	Yes	No	No	WA, GC	1	0	No	Div. of Water Resources
New Mexico	A	Yes	Yes	No	Yes	GC	8	3	Yes	State Engineer's Office
New York	R	Yes	Yes	No	Yes	GC	3	1	No	Dept. of Environmental Conservation
North Carolina	R	Yes	Yes	No	Yes	GC	0	0	No	Div. of Water Resources
North Dakota	A	Yes	Yes	No	Yes	GC	1	1	Yes	State Water Commission
Ohio	R	Yes	Yes	No	No	GC	3	1	No	Dept. of Natural Resources
Oklahoma	A	Yes	Yes	No	Yes	WA, GC	3	0	Yes	Water Resources Board
Oregon	A	Yes	Yes	Yes	Yes	WA, SC	2	0	Yes	Dept. of Natural Resources
Pennsylvania	R	Yes	No	No	No	GC	5	2	No	Dept. of Environmental Resources
Rhode Island	R	No	No	No	No	GC	0	0	No	Water Resources Board
South Carolina	R	Yes	Yes	No	No	GC	0	0	Yes	Water Resources Commission
South Dakota	A	Yes	Yes	No	Yes	WA, GC	1	0	Yes	Dept. of Water & Natural Resources
Tennessee	R	No	No	No	Yes	GC	2	0	Yes	Dept. of Health & Environment
Texas	A	Yes	No	No	Yes	WA, GC	5	3	Yes	State Water Commission
Utah	A	Yes	Yes	No	Yes	WA, GC, SC	3	1	Yes	Dept. of Natural Resources
Vermont	R	No	No	No	Yes	GC	1	1	No	None
Virginia	R	No	Yes	No	No	GC	4	0	No	State Water Control Board
Washington	A	Yes	Yes	No	Yes	WA, GC	0	2	No	Dept. of Ecology
West Virginia	R	No	No	No	Yes	GC	3	0	No	Div. of Natural Resources
Wisconsin	R	Yes	Yes	No	Yes	WA, GC	3	2	Yes	Dept. of Natural Resources
Wyoming	A	Yes	Yes	No	Yes	GC	9	1	Yes	State Engineer's Office
Territories										
American Samoa	O	No	No	No	No	GC	0	0	No	Dept. of Public Works
Guam	R	No	No	No	Yes	GC	0	0	No	Environmental Protection Agency
N. Mariana Island	O	No	No	No	Yes	GC	0	0	No	None
Puerto Rico	O	Yes	Yes	No	No	WA	0	0	No	Dept. of Natural Resources
Virgin Islands	O	Yes	Yes	No	No	WA	0	0	No	Dept. of Public Works

* A—Appropriation; R—Riparian; O—Other
† WA—Water Agency; GC—General Courts; SC—Special Courts

As a result, none of the western states adopted the riparian doctrine although a number of them retained some of its features. At first the appropriation doctrine was considered only to apply to surface water abstraction but now also applies to groundwater.

A summary of the system of water rights in the USA is given in Table 9.1.

OTHER COUNTRIES

The legal systems described above apply in the USA, Britain and the Republic of Ireland. Other countries have their own systems of water ownership which are mainly covered by legislation. The systems of water rights for several different countries are described below to provide a brief overview of the diversity which exists world-wide. These notes are really only for general guidance and the appropriate authorities should be consulted at an early stage.

Denmark

The local authorities prepare plans for the public use of water resources and control abstractions by a licensing system. Any private abstractions from groundwater are controlled within the water supply areas but there is no restriction for householders outside these areas. Riparian owners may abstract water for stock watering. The local authorities (Communes) license all minor abstractions up to 3000 m^3 a year or larger quantities for the joint supply of up to four households, including agricultural use. All other abstractions are licensed by the county councils, including all abstractions for spray irrigation. Licences are all time limited with a 10 year limit on all surface water abstractions, a 15 year limit on groundwater abstractions for irrigation, and a 30 year limit in all other cases.

Germany

There is a licensing system which applies to all major abstractions and is administered by the Länder or state governments rather than the federal government. Interpretation varies from one state to another so early consultation is a good idea. Licence control does not extend to abstraction by landowners for their own or others' drinking, washing and watering; and in moderate quantities for watering stock and other uses on farms. Groundwater can be abstracted in moderate quantities for agriculture, forestry and horticulture. All major abstractions require licensing, as do all abstractions within defined protection zones.

France

In France, water abstraction is controlled by the local authorities (Communes) and by the Prefects who are outposted civil servants with responsibilities for certain legal aspects as part of the Ministère de l'Environnement.

The rights to Private Waters are very similar to water rights under English common law. Permits are required for some abstractions. For drinking water abstractions from surface waters it is necessary to notify the commune and for industrial or irrigation abstractions authorisation is needed from the Prefect. The Prefect also authorises all large groundwater abstractions. There are unlikely to be any restrictions for abstractions for domestic supplies although there are limits to the depth that boreholes can be drilled in some areas.

Italy

In Italy, waters are divided into two categories: Public Waters and Non-public Waters. Public Waters are all those rivers, streams, lakes and groundwater abstractions which are declared to be Public Waters by the Ministry of Public Works. All other waters are, by definition Non-public Waters.

There are restrictions on the use of Public Waters. These are controlled by the Ministry of Public Works at state level and the Regional Authorities at a local level. The rights to use Non-public Waters are much less restrictive and are very similar to those under English common law. It is unlikely that you will experience any serious legal problems in obtaining a domestic water supply in Italy. Consult the Ministry of Public Works to make sure that there are no restrictions and also the local authority.

Netherlands

The Rijkswaterstaat is part of the Ministry of Transport and Public Works and is responsible for water resources planning at national level for major surface waters. Other waters are controlled by local Water Boards which are made up of elected representatives. The Water Boards license abstractions.

Japan

Surface water abstraction has been controlled by legislation since 1961 with management being based on six major river basins. The permitted quantities for major abstracters are reviewed each year. There are complex long-existing rights for abstraction, particularly for agriculture, which may take precedence over the newer system. There is little control on groundwater abstraction except to limit the potential for subsidence caused by pumping from certain aquifers.

Singapore

The Public Utilities Board provides water supplies for the whole population and there are no private abstractions.

Spain

The Spanish legal system classifies waters into Public and Private Waters. Public Waters comprise all large streams and rivers, all water courses and springs on public land and groundwater beneath public land. Private Water includes streams which wholly flow on private land, groundwater beneath private land and rain-water which falls on and is collected on privately owned land.

The rights to use water are complex but, in general, the private abstraction of running water for drinking, stock watering and domestic purposes are exempt from control. Landowners may use springs and groundwater sources for up to 7000 m^3 a year. All other abstractions are subject to Basin Plans which preserve a proportion of the available water for public use.

Zimbabwe

All water is vested in the state and major surface water abstractions are controlled. There is no control over groundwater abstraction. Boreholes must be registered and in major urban centres permission is needed for drilling new wells. Other than that, there is no control over or record of groundwater use. Abstractions from surface waters are permitted by water courts who take into account the current use and available resources. At the moment the government is reviewing the water legislation and greater controls are likely in the longer term.

10
Problems with External Causes

From time to time, private water supplies can be affected by the activities of other people. Derogation is caused by someone else's abstraction reducing the quantity of water available to you and has already been mentioned in earlier chapters. The threat of pollution from broken sewers, septic tanks and cess pits has also been discussed. Unfortunately, there are a number of other ways in which your water supply can be threatened. The yield of a spring or borehole can be reduced by causes other than straightforward derogation and there are many ways in which water can become polluted other than from broken sewers. A few of the problems caused by the activities of other people are considered here, with hints on how to solve them.

LOSS OF YIELD

The first sign that a source has suffered a reduction in yield is often when the taps dry up. Of course, they may not dry up completely. The problem may just be that the storage tank takes longer to fill than it used to and perhaps you have to wait for your turn to have a bath.

If you are in the middle of a long hot dry summer the most likely cause is that there has not been sufficient rain. This sort of problem is often made worse because in such dry conditions water requirements often go up alarmingly. In such conditions there may be very little that can be done to solve the problem other than to deepen wells and boreholes or look for a new more reliable spring. Drought conditions are the ideal time to look for a reliable spring as you can select the one with the best flow during a drought!

If the supply fails and there is not a drought the cause is probably one of those listed below. You may get a clue as to the cause if the supply gradually deteriorated over a period of time rather than suddenly failing, although this may be difficult to notice in the early stages.

The Effect of Derogation

Derogation can affect every type of water source except rain-water collection tanks. If you rely on a stream intake, a new abstraction upstream may reduce the flow of the stream enough to prevent you obtaining sufficient water. In these circumstances check the legal position as outlined in Chapter 9 and then go and talk to your neighbour. You may be able to sort things out and avoid any drastic action. When a spring, well or borehole has been affected by a new abstraction, it is often difficult to prove and you may need expert advice.

If you live in a country where water abstraction is controlled it is always a good idea to check with the local office of the regulatory authority. They will take appropriate remedial action within the constraints of the laws and regulations. This is likely to mean that if the abstraction which affects your water source is either illegal or recently permitted, the regulatory authority will sort things out; otherwise you will be on your own. It is important to obtain good advice at this point. Remember the tips mentioned in Chapter 1 and make sure that any expert you employ really does have experience of the sort of problem you are facing. Do not be afraid to ask them for the names of a few previous clients and check out how these people think of them. These comments apply equally to lawyers as to technical experts.

Problems of loss of stream flow may also be caused by the construction of a new dam, which need not necessarily be a large affair but just a temporary dam made with a few boards or sand-bags to give someone deep enough water to pump from the stream.

If your stream periodically runs dry, have a walk along it and you will soon find the answer. Again you may be into a legal argument but try to sort things out by talking with your neighbour rather than taking them to court.

The remedy for derogation may be to find an alternative supply or to deepen your well. Where groundwater levels have been lowered in artesian aquifers, however, the usual answer will be to install a pump into the well. You are likely to be able to continue having a water supply but the costs will be increased. In some instances, however, the effects can be more complicated than they appear at first sight. An abstracter I know uses the artesian discharge from a small diameter (50 mm) tube well to supply his house and smallholding. During a pumping test on a newly constructed well nearby, the flow dried up and only returned to a trickle when the test pumping stopped. Investigations showed that the artesian pressure head had been restored suggesting that the tube well had become blocked. It was found that several metres of silt had accumulated in the bottom of the well and once this had been removed the flow was restored. Figure 10.1 shows how this silt was removed.

Civil Engineering Works and Quarries

There are quite a number of different ways in which the water table can be lowered when new buildings and roads are being built. When a water table is lowered, a spring may have a reduced flow or may dry up altogether. Falling water tables can also seriously reduce the yield from wells and boreholes. The root of the problem is usually that large holes are being dug and this is possibly made worse by water being pumped from them.

In Chapter 2 the concept of groundwater storage was discussed. In most aquifers the proportion of water in saturated rocks may be less than 1% and could be as much as 35% of the total volume. A large hole dug below the water table can be thought of as a sort of aquifer with a 100% storage. Even if all the water is left behind when the rock is taken out, there is not enough to fill the hole. The water level in the excavation will be lower than in the surrounding rock. Water will flow from the rock into the hole to equalise these levels. This, in turn, will cause water levels in the surrounding area

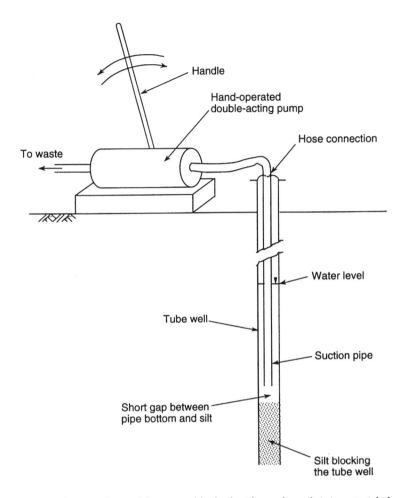

Handle

Hand-operated
double-acting pump

Hose connection

To waste

Water level

Tube well

Suction pipe

Short gap between
pipe bottom and silt

Silt blocking
the tube well

Figure 10.1 If your tube well becomes blocked with sand or silt it is a straightforward operation to clean it out. The method shown here is for the silt to be pumped out using a double-acting reciprocating hand pump with the suction extended by a length of suitable tubing such as 25 mm diameter plastic pipe. The effect is very similar to using the attachments on a domestic vacuum cleaner to reach the dust in an awkward corner. When you insert the suction tube, push it to the bottom until you can feel the top of the blockage. Then pull it back by about 100 mm and start pumping. It is important to leave this gap to prevent the pipe becoming clogged with the sediment. If the treatment is working, a mixture of water and sediment should be pumped out. Pump steadily until the water starts to clear and then lower the suction pipe following the same procedure as before. Continue until the water is clear and your probing with the suction pipe cannot find more sediment. Do not use a powered pump as the pumping action is very likely to be too strong and cause more sediment to be drawn into the tube well. An alternative method is to connect the suction pipe to a water supply and gently wash the sediment out. Take care to carry the operation out steadily in several stages, steadily flushing until the water clears each time. Do not try to flush it out using the full pressure of your water supply; this is likely to push the fine-grained material into the well face and make the problem much worse

to fall. Figure 10.2 illustrates this effect in the case of quarries or open pit mines.

In rocks where flow is through fissures and perhaps the storage value is less than 1% the large contrast in storage may be very significant and the effect can spread for some distance away from the hole. Springs can therefore dry up easily and the yield of wells or boreholes can be affected. Such excavations include trenches for new gas mains or water pipes, holes dug for the foundations of new buildings, and cuttings for new roads and quarries.

When an excavation is made any distance below the water table it is usually necessary for water to be pumped so that work can continue. This makes matters worse by extending both the duration of the effect and the area over which groundwater levels are lowered.

After pumping has ceased, the water level will slowly build up in the excavation until the original groundwater levels have been restored. Sometimes it can take two or three years before water levels have been built up again. Once this happens the yield of sources should be back to normal.

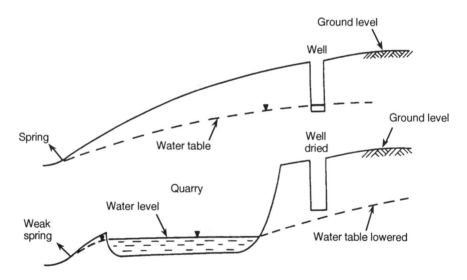

Figure 10.2 When quarries are dug below the water table they can cause permanent lowering of groundwater levels simply by being there. The top picture shows a water table sloping towards a spring following the ground topography. When a large quarry is dug it effectively moves the spring line to the quarry back wall and the water table is lowered in the area around the quarry. The well seen in our example becomes dry and the former spring is weakened or even ceases to flow after the quarry has been dug

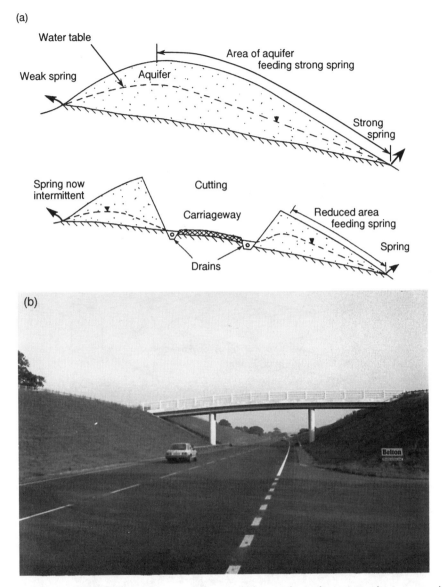

Figure 10.3 (a) A road cutting can divert groundwater from the surrounding area and cause the water table to be lowered, affecting local water supplies. In the top drawing a small aquifer is shown forming a hill. It contains a water table which slopes gently, mimicking the surface topography and forming two spring lines. When a road cutting is built through the hill the water table is lowered with water being diverted to the large drains placed on each side of the carriageway. This has the effect of weakening the flow of the springs, with one reduced to being intermittent. In one example I know, a farm water supply was badly disrupted by a road cutting to the extent that there was not enough water for the domestic needs of one person. The lowering of the water table was so great that the yield of a borehole was also badly depleted. (b) A road cutting which fully penetrates a shallow sand aquifer. The drains down each side of the road divert groundwater to the local streams. (Photograph: Rick Brassington)

(a)

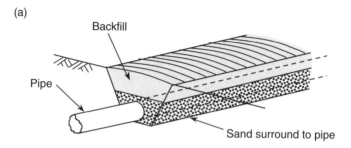

(b)

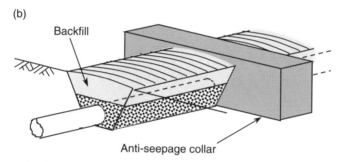

In some instances, such as with road or railway cuttings, this cannot happen. Drains are usually installed along the side of the road to prevent any build-up of water and the road flooding. As these drains will prevent any build-up of groundwater, the loss in yield will be permanent. Figure 10.3 shows the impact of cuttings on the local water system. Badly filled pipe trenches can also act as drains and have the same detrimental effect, as can deep tunnels. These effects are illustrated in Figure 10.4. Sadly, the impact on small water sources caused by pipes, roads and other similar constructions are rarely given consideration by the engineers who build them. If you are affected and complain, the most likely reaction is one of sheer disbelief. You may have to employ a hydrogeologist to investigate your problem and possibly a lawyer to argue your case before you are taken seriously.

When spring flows are reduced as a result of quarrying or other excavations, there is little which can be done to restore the yield. The only remedy may be an alternative supply, perhaps from a new borehole. Boreholes which have suffered a reduction in yield caused by a lowered water table can be improved by being deepened. Sometimes, however, simply lowering the pump may be enough to restore an adequate supply.

Land Drainage

In shallow aquifers the deepening of ditches or the installation of tile field drains will usually lower the water table in the immediate

---

Figure 10.4 Long pipelines are used to transport fuel oils, gas and chemicals and for long-distance water transfer. The pipes used are typically 0.5–2.0 m in diameter and require a large trench to accommodate them. It is a standard technique for the pipes to be bedded down in the trench on a layer of sand with the rest of the trench being back-filled with the material originally dug out as shown in (a). The sand and backfill can act as a highly permeable layer and effectively turn the trench into a land drain. Where the trench cuts across aquifers groundwater may drain along it thereby drying up springs and depleting the yield of wells. (b) Anti-seepage collars are easily constructed along the line of the trench which will prevent this flow. They comprise a short trench about five times as long as the pipe diameter, cut at right-angles to the main trench. They should extend below the base of the main trench to a depth of at least twice the pipe diameter. Once the pipe has been placed the cross trenches are simply filled with concrete to form a barrier to flow. In areas where such leakage is likely, the collars should be constructed at intervals roughly equal to ten times the pipe diameter. (c) A pipeline under construction. Notice the significant depth of the trench and the water standing in the trench bottom. (Photograph: Rick Brassington)

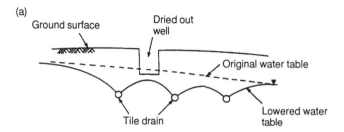

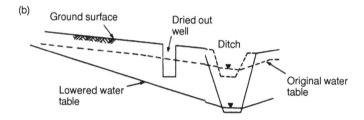

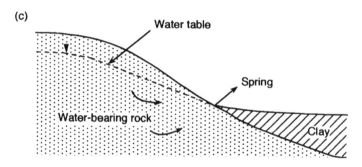

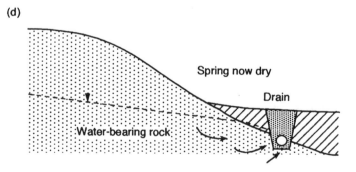

Figure 10.5

area; after all, that is what they are designed to do. When the aquifer is quite permeable (e.g. sand and gravel), the water table may be lowered over a wide area and shallow groundwater supplies may be affected. Figure 10.5 shows a few examples of the different ways in which this problem can occur. The only remedial work which can be undertaken where land drainage affects a source is also to replace the source. In this case though, provided that the water quality is good, it may be possible to use the water collected in the drains as the source of supply.

Lack of Maintenance

The most usual case of failure of any water source is lack of maintenance. Storage tanks and pipes can gradually start to leak. This may not be easily spotted except during careful routine maintenance checks. Slow leaking from the system can divert very large volumes of water into the ground, so if you have a water supply failure check for leaks straightaway.

Slow silting up of pipework can also have the same effect, resulting in a reduced flow out of the taps. Where silt builds up in the pipe system there may also be a quality problem caused by organic material in the silt starting to rot. All types of sources are subject to this silting problem, which again can be avoided by regular routine maintenance.

Pump wear tends to be a gradual process and results in a reduced yield. Quite often the only way to discover if this is the cause of your problem is to remove the pump and examine it on your workshop bench. If the impellers appear to be in good condition and the bearings are not worn then look for another cause of your loss of yield.

Figure 10.5 When land is improved by new ditches and drains being installed, the water table will be lowered. After all, that is what the drains are designed to do. The top drawing (a) shows how tile drainage pipes modify the shape of the water table and may affect the water level in a local well. A similar effect is seen in (b) caused by a ditch being deepened. Here, the water table now flows to a lower base line, causing levels to fall generally. The example in (c) shows a spring line thrown out where clay butts up to a permeable rock. After the clay field has been improved by drains being installed (d), the groundwater flow is diverted to the drain, drying the spring up completely

There are two special cases that deserve particular attention which involve clogging of the face of a well or borehole and the gradual deterioration of a spring supply. The problem of clogging wells and boreholes has already been described in Chapters 3 and 8. It is worth remembering that gradual clogging of a well face and the consequential loss of yield is often blamed on other causes. There have been many instances where accusations of derogation have been made very hotly indeed but close examination of the problem has revealed that the loss in yield has been caused by a deterioration of the well face. Do not forget to include specific capacity measurements as a routine part of your maintenance programme if you use a well or borehole.

Slow deterioration of the yield of a spring can be caused by a gradual build-up of silt or chemical deposition which prevents the water getting into your catch pit. The spring will still be operating but will be breaking out at a different place on the hillside. The signs to look for are a build-up in the catchpit chamber and the development of boggy ground around the spring source. The remedial action is to reconstruct the spring collection chamber completely. Do not be afraid that you will lose the water by digging into the hillside. If your catch pit is the lowest possible discharge point for the aquifer in the immediate area, you are bound to collect all the available spring water.

DETERIORATION IN QUALITY

Water pollution problems can be caused by the way in which householders get rid of their waste materials. If your water supply is suffering from quality problems, first make sure that you are not the cause. Better still, adopt practices which will not cause such problems in the first place.

Many quality problems can be traced to a leakage of sewage. Always make sure that a water source is at least 30 m away from any septic tank, cess pit or sewer pipe. If you use water from a stream, it is likely to be subject to pollution from sewage discharging into it. This may only be the overflow from a septic tank but can cause just as many problems.

In some instances, pollution of a well by sewage can result from the cleaning out of a ditch. Over the years the accumulation of mud and silt in the bottom of the ditch will tend to seal it off from an aquifer. This will mean that if sewage discharges are made into the stream or ditch, none of the contaminated water can flow into the aquifer. If the ditch is cleaned out, however, the polluted water may easily get into the aquifer and pollute nearby wells. Similar risks of pollution from industrial discharges into watercourses also exist.

Remedial action is often to construct a completely new source, usually a borehole. It is often very difficult to attempt to clean out polluted sources, like shallow wells, as the contaminated water may linger in the ground for a considerable period of time. Polluted streams, on the other hand, will quickly recover as the pollution will flow quickly downstream with the water, but you will need to disinfect your stream intake before it can be used again. Perhaps you should consider a new source altogether if this is likely to be a frequently occurring problem.

SEWAGE TREATMENT

The 1980s were designated the "water decade" by UNESCO as part of a programme designed to ensure that everyone on the planet would have an adequate, clean and reliable water supply at the end of it. Sadly the task proved too great to be solved on that timescale but work is continuing. Much more effort has been put into providing sewage disposal systems than into finding new water sources because it was found that disease from polluted water has a greater impact than simply a lack of water. A significant proportion of contamination problems associated with private water supplies is caused by sewage, which somewhat ironically, usually originates from the same household. You can do a couple of things to avoid this problem; first ensure that your sewage is properly treated and, secondly, ensure that your water source and sewage treatment unit are far apart. Some years ago I visited a cottage in Shropshire as part of a well survey exercise. It had been modernised and extended by two cottages being knocked into one. One of the two wells was no longer needed so it had been converted into a cesspool, polluting the remaining well which had grown a bright green crust. You will not be surprised to learn that I declined the offer of a cup of tea!

Cesspools and Septic Tanks

In most western countries the sewage system uses water to flush the lavatory and wash the solids along the sewer pipes. Where a public sewer is not available the alternatives are cesspools or septic tanks. Many people think that these are different terms for the same thing but

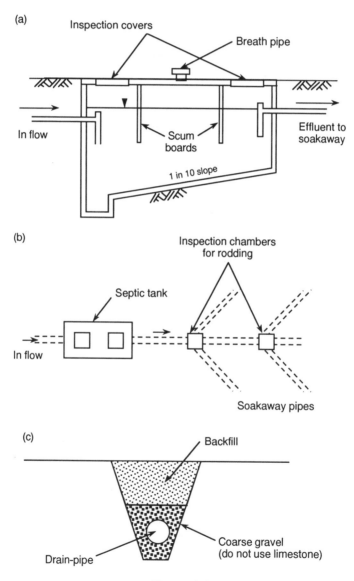

Figure 10.6

this is far from the case. A *cesspool* (or cesspit) is an underground chamber which is used simply to hold sewage before it is taken away by tanker for disposal elsewhere. A *septic tank* on the other hand, provides a sewage treatment process using naturally occurring bacteria.

It is difficult to see any great advantage in using a cesspool to dispose of your sewage which needs frequent emptying, when you bear in mind that a septic tank needs very little attention and will operate successfully for many years. Septic tanks were first developed in the middle of the 19th century in France. The system was later introduced into Britain where it was patented, and then into the USA by 1883. The design has been modified and improved a number of times since then and recently has been produced by several companies in a ready-to-install form made from glass fibre and other plastic materials. There are now a very large number of septic tanks in use around the world. For example, there are estimated to be more than 25 million in the USA alone.

A septic tank installation consists of two essential elements: the septic tank and a soakaway system for effluent disposal. The septic tank is a buried watertight container which is designed to provide anaerobic digestion for the organic matter in domestic sewage. The digestion process produces a solid fraction which is retained in the septic tank and a supernatant liquor which flows out of the tank for further treatment in the surrounding environment. Ideally, the effluent flows through a percolation distribution system into the soil where there is a huge capacity for natural treatment by the bacteria which live in the soil. In some instances the supernatant flows into a soakaway pit which may not be capable of receiving the higher rates of flow, or is discharged directly into a stream where there is a high risk of pollution. The main features of a septic tank and the associated percolation system are shown in Figure 10.6.

Figure 10.6 (a) Shows a longitudinal section of a septic tank which is divided into three chambers by scum boards hung from the top to prevent floating material from moving through the tank. The sewage flows into the first chamber where the solid matter starts to break down. Bacterial activity digests the organic material as the sewage flows through the tank. Sludge is formed which falls to the bottom of the tank and collects in the sump. It should be removed on a regular basis. Only water is discharged from the outlet pipe, although it has a polluting content. The best method for disposing of this effluent (and in some areas the only one permitted) is through a series of porous drain pipes as shown in plan form in (b). Make sure that there is a fall of at least 1 in 40 and build inspection chambers at the junction points. You will be able to rod the pipes to clear any blockage through these chambers. The design of the soakaway drains is shown in (c). Use unglazed clay or perforated plastic pipes, leaving gaps at the joints and surrounding the pipes with coarse gravel. Do not use limestone as this will eventually fuse into an impermeable mass

When you first put your newly constructed septic tank into operation, fill it with water before running sewage into it. It will take a few days before the bacterial processes are firmly established and the tank is treating your effluent.

The prime function of a septic tank is as a settlement chamber with only a limited treatment capability. The sludge which settles on the floor of the tank is partly digested by micro-organisms which produce various gases as a by-product consisting of mainly carbon dioxide and some methane. The fats, oils, greases and soaps in the sewage are brought to the surface by these gases to form a thick scum which provides a good indication that the septic tank is working properly.

The components of domestic waste-water have changed over the years, reflecting the use of different materials in the home. The main change has been the introduction of detergents, and concern for the effects that these chemicals may have on the biological processes in septic tanks has led to kitchen and bathroom waste-water being piped to a separate soakaway system. It is now generally thought that the average concentrations of household detergents and cleaning materials in domestic sewage do not adversely affect the functioning of septic tanks. There are a few things which you should remember *not* to do as they could ruin your septic tank's digestive system, such as not pouring old paint, thinners or fuel oil down the drain. You should dispose of all of these materials in a properly licensed waste disposal facility or else burn them carefully in a controlled way.

Although septic tanks are likely to be trouble-free for years, it does not mean that there is no need to maintain them. It is generally recommended that the sludge is removed about once a year for the average household. Do not empty the tank completely; it is possible to pump the sludge out separately.

Non-water-borne Sewage Treatment

In the rural parts of many countries where water supplies are limited, people produce very little sewage. In such circumstances the old-fashioned "earth closet" was found at the bottom of each cottage garden. These consisted of little more than a deep hole, over which a small wooden hut with a seat was set. In many instances these toilets were smelly and unpleasant to use. Pit latrines which do

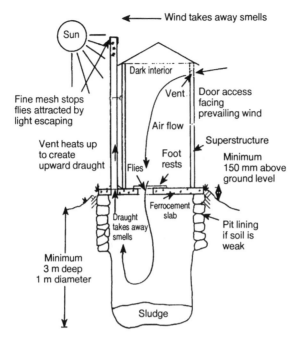

Figure 10.7 Ventilated improved pit latrines have made very significant improvements in preventing the spread of disease in areas of the world where water-borne sewage systems are not practical. The ventilation is provided by a pipe connected to the pit and taken to a height above the roof level. The vent pipe heats up in the sun causing an upward air flow, taking away the smell and keeping any flies inside the pit. (Reproduced by permission of WaterAid, London)

not require any water have been developed so that they do not have a smell problem. Figure 10.7 shows the main principles of construction and operation of ventilated improved pit latrines of a type which is now in widespread use in many developing countries but are suitable for use anywhere.

STORAGE TANKS

Fuel oils and many other liquid chemicals are stored in tanks which, if they leak, can cause serious water pollution problems. These problems are made much more difficult to resolve where the tanks are buried underground as it is difficult to detect when they are leaking. There are very many such tanks. For example, there are between five and six million in the USA, of which 70% are for farming or non-commercial use. The alarming thing is that the US EPA estimate that about 400 000 of these tanks are leaking.

If possible, tanks should not be buried but set on firm supports and surrounded by a bund or wall which is capable of retaining the tank contents should it leak. Modern underground tank installations require special precautions to limit leakage and enable any leak which does occur to be detected. If you are installing tanks for fuel storage check with your local council or environmental regulator who will provide you with details of the precautions they require.

If an underground tank leaks it should be dug up and removed to an approved disposal site. The contaminated soil is also usually dug up and taken to a landfill. In some cases it is possible for the soil to be treated by incineration or washing. If your water supply has been affected by a leaking tank it will be necessary to remove the tank before you can start to think about cleaning up the source. If you take water from a stream then it may simply be a matter of waiting until the contamination flows past your intake. Groundwater supplies are likely to require long-term pumping to remove the pollutant from the aquifer.

ACCIDENTAL SPILLAGE

It is quite surprising how much polluting material is carried around the country in tankers. Each time you make a journey on a major road you will see a large number of tankers carrying a wide variety of liquids ranging from fuel oils to chemicals and from milk to hazardous wastes. Fortunately, relatively few accidents occur where tanker loads of material can spill into streams or seep into aquifers, but these things do sometimes happen. The main threat is to streams and shallow groundwater sources. The potential problems arise from the spilt liquids flowing into streams and from them being washed into the stream as the emergency services hose the spilt liquid away. These days care is usually taken to reduce the impact of the clean-up operation but a certain amount of damage is unavoidable.

There is very little you can do to prevent this problem, but you can predict its likelihood by considering the proximity of roads to your source. Do not be fooled into thinking that tankers never go down narrow country lanes, as practically every farm receives deliveries of fuel for the tractor, etc. If a spillage occurs then the local pollution prevention authority will spring into action. You should contact them and explain your concern about your water supply and they will advise you of any risks and take your supply needs into account when they plan how to clean up the mess. Once more the only solution may be to look for a new source of water which again will probably turn out to be a borehole.

SILAGE LIQUOR

Modern agricultural practice tends to use grass to produce cattle feed in the form of silage rather than the more traditional hay. Silage is produced by finely cut green plants (mainly grass) being tightly packed into tanks (silos) and fermented. This process produces a dark highly polluting liquor which should be treated and disposed of in a safe manner. Pollution of springs and wells by silage effluent is a common problem in rural areas, especially where the main aquifer is a fractured rock suck as limestone. If your well is polluted, stop using it at once and arrange for a temporary supply. Locate the source of the problem and take measures to stop further leaks or discharges. Pump your well to waste to draw the pollutant out of the aquifer, taking care where you discharge the pumped water.

It is always a good idea to contact the pollution prevention authority who are likely to help locate the source of the problem and provide advice on the best place to pump the waste water. It may take several weeks to clean up the aquifer. Disinfect the well and all the pipe system before you use the well again. If this does not work then you will probably have to drill a new well. Do not simply abandon your well and leave the silage effluent to leak into the ground. It will cause a greater problem and may eventually affect your new well and those of your neighbours.

NITRATES AND PESTICIDES

Modern agricultural practice involves the use of a host of chemicals, many of which get into the hydrological cycle. Over the past few years the problem of nitrates in water supplies has received a great deal of publicity and has been the subject of a lot of close scrutiny. Nitrate in the diet is harmful to health and can kill babies under three months old if taken in sufficiently large concentrations. This is one of the causes of a condition called *methaemoglobinemia*, and in this case is brought about by molecules of nitrogen taking the place of oxygen in the blood. As the body cannot use nitrogen, there is no way in which these molecules can be removed. With nitrogen in the way there is no room for oxygen molecules to be carried round the body in the bloodstream and so severe oxygen starvation sets in. In young infants this can quickly cause death and can only be treated by total replacement of the blood. Some medical authorities believe that, although methaemoglobinemia will not kill older children and adults, it can have a significant medical effect. The maximum permitted nitrate concentration in drinking water is 50 mg/l and takes this into account.

Over the past couple of decades there have been a huge number of chemical pesticides produced. These are designed to kill insects and plants and if not used correctly can cause serious pollution problems. They fall into two categories: persistent chlorinated chemicals such as DDT, endrin and aldrin; and non-persistent but toxic pesticides which are used in great quantities by farmers in all Western countries. The environmental regulators in these countries have shown that small concentrations of these chemicals can be found in almost all waters in the natural environment. If you are farming in the catchment area of any water supply you should take care to ensure that the chemicals you use and the way that you use them will not contaminate the water.

Nitrate gets into water supplies by percolation from agricultural land into aquifers. In areas where farmers use weedkillers and pesticides in any significant quantity these two can percolate into aquifers and pollute groundwater resources. If you live in an arable area and your water supply is from a well or spring, you would be well advised to periodically have samples analysed for nitrate and pesticide content. If your source is affected, the best remedy is, once again, to construct a new deep borehole as shown in Figure 10.8. The lining of the boreholes should be designed so that water can only get into the borehole at some depth below the water table. This means that the majority of water entering the borehole will be old groundwater which will dilute any nitrate-bearing groundwater which has recently entered the aquifer.

Where a water supply does not come within the drinking water quality limits, many of the problems can probably be solved by overhauling the system following the guidelines set out in this book.

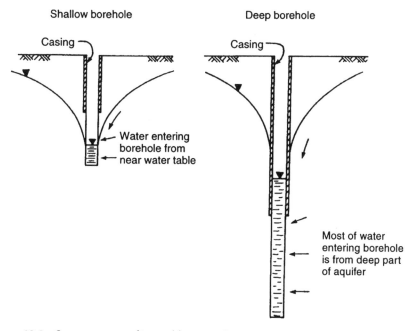

Figure 10.8 Some water quality problems, such as increasing nitrate levels, cannot be so easily solved and are likely to require a new water source. If the affected source is a spring, land drain or shallow well, replacement with a borehole which taps a deep aquifer will probably do the trick. The diagram shows how deepening a borehole attracts water from a greater depth. this will reduce the concentration of nitrate or other contaminants which originate from the surface. Sadly, however, the solution may only be temporary with nitrate concentrations starting to creep up again after only a few years

It is possible that many people with this sort of problem will be pushed towards seeking a mains water supply. This could well be the only answer in some cases, but do not be in too much of a hurry to accept it as the only one. If a new or deeper borehole is likely to provide a satisfactory supply it could be a great deal cheaper than a supply from the public main.

QUALITY PROBLEMS WITH QUARRIES AND MINES

In areas where there has been a lot of quarrying there may be a large number of flooded quarries. These often present a particular risk. Water in these quarries is usually groundwater and so any polluting material which enters the quarry immediately pollutes the groundwater. This polluted water flows easily into the aquifer and may pollute wells, boreholes and springs which are located close by.

Mining can produce water quality problems caused by the minerals being worked or those which are found associated with the main ore. Unless properly carried out, metalliferous mining can cause problems as water enriched with dissolved metals flows into local streams. It would be wise to have your water supply tested if you abstract downstream of such mines.

Coal mining can also cause serious water quality problems. Iron pyrites is a mineral commonly found in coal-bearing rocks. When it is exposed to oxygen-rich waters a chemical reaction takes place which produces sulphuric acid and iron oxides. The iron oxides are not soluble in water and precipitate to form an orange-coloured sludge. Unless this happens exactly where you abstract your water it is not likely to cause a problem, unless lumps of the sludge are washed into your stream intake. The acid may result in more problems, however, as it may affect the quality of your supply and require a greater level of treatment. Groundwater pollution is unlikely from these sources except where water pumped from the mine is allowed to soak into the ground. There are a number of well-documented cases of serious long-term pollution being caused in this way.

AIR-BORNE POLLUTION

The effects of burning fossil fuels on atmospheric pollution are well known. The acidic pollution of Norwegian lakes caused by coal-burning power stations in the UK and Germany has been established. The results of a research project in the Pennines of northern England, mentioned in Chapter 6, showed that acid rain and air-borne pollution from motor vehicles have a similar impact on small-scale spring supplies. There is not much you can do about this type of problem except to monitor the situation regularly and take steps to ensure that surface water cannot enter the spring catchpit. After that the choice is either to treat the water or replace the spring with a borehole.

LANDFILLS

Waste disposal sites, whether they receive domestic waste or the more hazardous wastes produced by heavy industry, can cause serious water pollution problems. These days landfills in most Western countries require a licence to operate. Before a site licence is granted the regulator will take into account the threat of water pollution, which usually includes the local water abstractions. The risk of serious pollution from waste disposal sites is thereby reduced to a minimum. Completed landfill sites, however, remain a threat for some years after they have been abandoned, especially as these old landfills were not operated to modern standards.

The problems caused by polluted water flowing out of domestic waste disposal sites are very similar to those caused by sewage, although this polluted water can be much more concentrated. Pollution problems from industrial waste sites depend on the nature of the material that is being put there, but is usually similar to domestic refuse dumps. The cure for polluted supplies is usually to find a new source of water, especially where groundwater is involved. It can take many years before polluting material flows out of an aquifer.

RUN-OFF FROM ROADS

An unusual pollution problem can be caused by the run-off from roads picking up a variety of material and carrying it into the local water courses. The types of polluting material involved are salt during the winter, spilt fuel and lubricating oil, and a greasy deposit from tyres. These materials are only likely to be in sufficient quantities where a long section of heavily used road, such as a motorway, drains to a particular point. Motorway drainage is designed to cope with very heavy rainfall and consists of drainage pipes buried in a trench back-filled with gravel. These drains run by the side of the carriageway towards local streams but in some areas they are connected to soakaway systems. Water sources can be polluted but usually only if they are within fairly close proximity to motorways. Streams are mainly at risk with the threat to groundwater being minimal unless the drainage is diverted to soakaways.

Remedial action is either to change your source of water or divert the road run-off somewhere else.

Appendix 1
Conversion Factors

The units of measurement used throughout this book are the SI (Système Internationale d'Unités) or metric system which is based on the metre, kilogram and second. Any deviation from this system in the text has been made clear. The conversion factors provided in this appendix will allow you to convert from the metric system to either the American or British system of units based on feet, inches, gallons, pints and pounds. Please bear in mind that there are some differences between the size of some of the units used in the two countries.

From	To	Multiply by	From	To	Multiply by
		Length			
cm	in	2.540	in	cm	0.394
m	ft	3.281	ft	m	0.305
m	yd	1.094	yd	m	0.914
km	mile	1.609	mile	km	0.622

From	To	Multiply by	From	To	Multiply by
Area					
m^2	sq ft	10.764	sq ft	m^2	0.093
m^2	sq yd	1.196	sq yd	m^2	0.836
ha	acre	2.471	acre	ha	0.405
km^2	sq mile	0.386	sq mile	km^2	2.590
Volume					
litre	pint (UK)	1.76	pint (UK)	litre	0.568
litre	pint (US)	2.11	pint (US)	litre	0.473
litre	gal (UK)	0.220	gal (UK)	litre	4.546
litre	gal (US)	0.264	gal (US)	litre	3.785
m^3	gal (UK)	220	gal (UK)	m^3	0.0045
m^3	gal (US)	264	gal (US)	m^3	0.0038
m^3	cu ft	0.028	cu ft	m^3	35.32
m^3	cu yd	0.764	cu yd	m^3	1.309
Flow rates					
l/s	gal/min (UK)	13.2	gal/min (UK)	l/s	0.076
l/s	gal/min (US)	15.85	gal/min (US)	l/s	0.063
m^3/day	mgd (UK)	220	mgd (UK)	m^3/day	0.0045
m^3/day	mgd (US)	264	mgd (US)	m^3/day	0.0038
Abbreviations					
cm	centimetre		in	inch	
m	metre		ft	foot	
km	kilometre		yd	yard	
ha	hectare		gal	gallon	
l	litre		s	second	

Appendix 2
Safe Working Practices

Many of the activities which are associated with building and maintaining your own water supply have some sort of inherent danger. Remember the old safety adage that you are your own safety officer; think through the safety aspects of each activity *before* you start and do not take risks. It is absolutely vital that you wear the right sort of protective clothing, test that there is a breathable atmosphere in any confined space you are planning to climb into, and to have someone to help and be available to pull you out if you get into difficulties.

I have tried to provide you with a few guidelines about safe working practices both in the main text and in this appendix. Please read them all and after that it is up to *you* to ensure that you have the proper equipment and training, and follow safe practices which are appropriate for the work you are undertaking, the local conditions and to meet any laws or regulations.

Many countries have legislation which covers safety at work. In addition, many government and local government departments, trade associations and other organisations involved with water supply work have their own safety codes. This information is likely to be available from local libraries, water departments or government offices.

Some tasks are dangerous if they are undertaken by someone working on their own, especially where it involves entering a confined space or working in a remote place. Do not take risks by entering confined spaces on your own. If you are in a remote place and cannot get someone to work with you, at least leave details with a reliable person so that they can arrange a rescue if you fail to turn up at an agreed time. Do not forget to let them know when you have finished, to avert any unnecessary alert.

Some of the projects will involve the use of specialist equipment. Where this can be hired take advice on its use from the hire company people. If you are improvising, for example with air-lift pumping, please remember that you are an amateur and need to take extra care.

CLOTHING AND EQUIPMENT

Wear a hard hat (safety helmet) with a chin strap if you are working below ground level or when you have other people working above you. Make sure that you get an approved safety helmet from an industrial clothing supplier and do not improvise with your Dad's old army helmet. The right equipment is not expensive and is a lot cheaper than a new head! Do not be tempted to take your helmet off because it is hot. It is there to protect you and can only do that job if it is on your head.

It is a good idea to wear boots or shoes with protective toe caps at all times when you are working. Even if you are doing nothing more dangerous than knocking in nails, they will save you a bruised toe if you drop the hammer. When you are digging or handling heavy equipment or materials such as concrete pipes, they are vital.

Protective goggles are necessary when you are using a power drill and for any other operation which may cause pieces to fly into your face. They are also essential when you are handling chemicals, including mixing a chlorine solution to use as a disinfectant. Gloves are also needed when handling rough materials or to prevent your skin from contacting contaminated water or strong chemicals and disinfectants.

Your personal safety equipment should include a first aid kit and I also suggest that you have some first aid training so that in an emergency you know the right thing to do.

CONFINED SPACES AND OPEN EXCAVATION

The definition of a confined space is very wide and embraces any enclosed area. In the context of private water supplies it includes such things as wells, spring catchpits, well-head chambers, tanks and covered reservoirs. Before entering any confined space look around it and decide if you will be able to get out again easily. Also give consideration to a breathable atmosphere and the likelihood of anything falling on top of you when you are inside. Before putting your hand inside, bear in mind that underground chambers and the dark corners around well-heads may be home to snakes or poisonous insects. Other potential hazards include trapping hands or feet when removing heavy covers and the danger of back strain.

There are three main hazards in terms of the gases which may be present in a confined space. These are explosive gases such as methane; toxic gases such as hydrogen sulphide; and an oxygen-deficient atmosphere which is rich in carbon dioxide, for example. Natural methane may be present from the decay of organic materials such as peat. It is the cause of the "will-o'-the-wisp", flickering lights which are sometimes seen in marshy areas. Methane can also be present in coal-bearing rocks and is the main component in "fire-damp", the explosive gas in coal mines. Methane can be detected by special electronic instruments or the more old-fashioned miner's safety lamp where the presence of the methane makes the flame flare.

Hydrogen sulphide smells of rotten eggs and can be produced by the decay of organic materials. It is commonly found in sewers and septic tanks, and can be associated with polluted water. It is very poisonous if you breath it in for any length of time, with the added hazard that in toxic concentrations you can no longer smell it. It is heavier than air so may accumulate in underground tanks and chambers. If you have hydrogen sulphide present make sure that you ventilate the confined space thoroughly.

Carbon monoxide is another toxic gas which may be present. It is produced by the incomplete burning of organic materials and is always present in the exhaust gases from engines. It is about as heavy as air but under certain conditions could collect in under-ground chambers, especially if you have been using a compressor or other equipment near-by. It is very toxic and kills easily.

Carbon dioxide is naturally present in the atmosphere. Its main hazard is that it is heavier than air and can collect in underground tanks and wells. It is the most common hazard caused by gases that you are likely to face and can be detected by lowering a lighted candle into the space. If the flame flickers or goes out assume that the atmosphere is oxygen-deficient and ventilate it. Do not use a naked flame if methane may be present.

When working round the top of wells, boreholes or shafts which are more than 350 mm in diameter, either reduce the size of the opening to prevent any risk of you slipping and falling in, or wear a safety-harness with a lifeline attached to a secure point. In most cases a lifeline should be short enough to prevent a free fall of more than 3 m, and be secured to an immovable object or held by at least two "top men". If you are going to use a ladder or step-irons, make sure that they are secure and in good condition before you trust your weight on them.

Before you carry out any work over chambers, wells or tanks, carry out a visual examination of cover plates, staging, etc. Beware of any corroded metalwork or rotten and/or slippery timber. Remember that apparently sound timber can be rotten underneath. If in doubt, do not proceed before carrying out any necessary repairs, or use a safety-harness and lifeline.

Various gases may be present in enclosed chambers, especially when they are in contact with an open well or borehole. These include carbon dioxide, which can cause suffocation; hydrogen sulphide, which may be highly toxic in certain concentrations; and methane, which is potentially explosive. Take precautions using a gas detector or miner's safety lamp to test the amount of oxygen and for the presence of explosive gases. If in doubt, do not enter. Barometric conditions are important as atmospheric pressure strongly influences the emission of gases. It is usually safe to enter a chamber if the barometric pressure is rising, as air will be entering the chamber. Carbon dioxide is heavier than oxygen, however, and may accumulate at the bottom of a chamber so still test for oxygen levels. If the pressure is falling it is likely to be dangerous to enter the chamber as gases may be flowing into it. If in doubt, do not enter without providing an air supply.

It is dangerous to enter any excavation which is more than 1.25 m deep without the sides being supported. The sides may quickly collapse causing burial and death or injury by crushing.

ELECTRICAL EQUIPMENT

If you are using a power drill or other electrical tools make sure that you are properly protected by a circuit-breaker. Ideally use a low voltage supply (110 volts), which will mean having the correct power tools. These days many power tools are "cable-less" and work on rechargable batteries. These tools are a good way to avoid safety hazards from high voltage electricity. Take care that you use equipment that will not create a spark if you are working anywhere near explosive gases.

HYGIENE PRECAUTIONS

When you are working around any water source you must take great care not to cause contamination. Avoid knocking dirt into a well or spring catchpit when you remove the cover. To do this, brush any loose material off before moving the cover at all. Any tools or equipment which you may use for cleaning or maintenance work should be thoroughly clean. If in any doubt disinfect it by washing it in a chlorine solution with a strength of about 20 mg/l of chlorine. Do not just dip it in the disinfectant; give it a good scrub and then rinse it in clean water. Follow the advice given in Chapter 6 regarding making up this solution. Remember not to use "pine" disinfectants. These are based on phenols, which are easily detected by taste and only very small quantities will contaminate a well or catchpit.

If you are going to put your hands in the water try to use rubber gloves (clean ones, of course!) or at least give your hands a thorough wash. If you need to enter the well or catchpit or any other part of your water supply system, besides taking precautions for your own safety make sure that you are wearing clean overalls and that your boots are clean. It is a good idea to scrub your boots in a chlorine solution before you get in to make absolutely sure.

Limit the tools which you take to the water source to keep the risk of contamination to a minimum. If you use any for maintaining your sewage treatment — pipe rods for instance — do not use them on your water supply. You may think that this point is too obvious to make, but believe me, people actually do this sort of thing.

Once you have completed your work it is a good precaution to disinfect that part of the system following the recommendations set out in Chapter 6.

Reading List

Anon. (1985) *Guide-lines for Drinking-Water. Volume 2: Health Criteria and other Supporting Information Supplies.* World Health Organization, Geneva.

Anon. (1985) *Guide-lines for Drinking Water. Volume 3: Drinking-Water Quality Control in Small-Community Supplies.* World Health Organization, Geneva.

Anon. (1987) *Design of Small Dams* (third edition). Bureau of Reclamation, US Department of the Interior, US Government Printing Office, Denver, Colorado.

Anon. (1991) *Water Treatment Handbook* (sixth edition). Degrémont Rueil-Malmaison Cedix, France.

Anon. (1993) *Guide-lines for Drinking-Water Quality. Volume 1: Recommendations* (second edition). World Health Organization, Geneva.

Anon. (1993) *Manual on Treatment of Private Water Supplies.* Drinking Water Inspectorate, Department of Environment, HMSO, London.

Brandon, T.W. (ed) (1986) *Groundwater: Occurrence, Development and Protection.* Water Practice Manual No. 5, Institution of Water & Environmental Management, London.

Brassington, R. (1988) *Field Hydrogeology,* John Wiley & Sons, Chichester.

Clark, L. (1988) *The Field Guide to Water Wells and Boreholes.* John Wiley & Sons, Chichester.

Daly, D. (1985) *Groundwater Quality and Pollution,* Geological Survey of Ireland Information Circular 85/1, Ministry of Energy, Dublin.

Daly, D., Thorn, R. & Henry, H. (1993) *Septic Tank Systems and Groundwater in Ireland.* Geological Survey of Ireland Report RS 93/1, Department of Energy, Dublin.

Driscoll, F.G. (1986) *Ground Water and Wells* (second edition). Johnson Division, St Paul, Minnesota.

Freeze, R.A. & Cherry, J.A. (1979) *Groundwater.* Prentice-Hall, Englewood Cliffs, New Jersey.

Jordan, J.T. (1980) *A Handbook of Gravity-Flow Water Systems for Small Communities.* Intermediate Technology Publications, London.

Gray, N.F. (1994) *Drinking Water Quality.* John Wiley & Sons, Chichester.

Todd, D.K. (1980) *Ground Water Hydrology.* John Wiley & Sons, New York.

Twort, A.C., Law, F.M., Crowley, F.W. & Ratnayaka, D.D. (1994) *Water Supply* (fourth edition). Edward Arnold, London.

Watt, S.B. & Wood, W.E. (1979) *Hand Dug Wells* (second edition). Intermediate Technology Publications, London.

Wright, G.R. (1985) *Pumping Tests.* Geological Survey of Ireland Information Circular 85/2, Ministry of Energy, Dublin.

Wright, K.R. (1990) *Water Rights in the Fifty States and Territories.* American Water Works Association, Denver, Colorado.

Glossary of Technical Terms

Abney level: a hand instrument for taking angles and levels along steep slopes.

acid: having a pH of less than 7.0; see pH value.

acid treatment: using acid to remove deposits from the face of a borehole or well to increase the yield.

air-lift pump: a means of lifting water from a well using compressed air to aerate the water so that the air–water mixture will rise to the surface.

air vessel: a small chamber fixed to the discharge side of a hydraulic ram. The chamber contains air which acts as a cushion absorbing the shock caused by pressure changes in the water.

alkaline: having a pH value above 7.0; see pH value.

annular space: the space or cavity between the outside of the borehole casing and the surrounding ground.

appropriation doctrine: a system of water rights in the USA which protects existing abstracters from the impact of new ones.

aquiclude: an impervious rock which does not transmit water.

aquifer: a porous rock which holds and transmits water. It yields useful quantities of water to wells and boreholes.

aquifer test: a pumping test designed to determine the long-term yield of a new well and its impact on local springs, etc.

aquitard: a poorly permeable rock which transmits some water but not in useful quantities.

artesian aquifer: an aquifer in which the groundwater is under pressure and confined beneath impermeable rocks.

artesian borehole/well: a borehole/well which penetrates an artesian aquifer and in which the water level rises above the top of the aquifer. Sometimes the water may flow out at the surface.

atmosphere: the gaseous matter which surrounds the earth. It is largely nitrogen and oxygen with traces of carbon dioxide, argon, helium and other inert gases. The atmosphere also contains variable amounts of water vapour and dust which are significant in determining weather conditions.

bacteria: a group of microscopic single or multi-cellular organisms. Some types cause disease in animals and humans.

baffles: a series of wooden planks or slats, metal plates or even a perforated plate, which is placed in a weir tank to remove turbulent flow.

bailer: a tool used in drilling for removing sludge, mud or sand from a borehole. It consists of a steel tube with an open end, the other end being either closed or fitted with a simple valve.

base flow: part of a stream flow made up of groundwater. It sustains the stream during dry periods.

bonded gravel screen: a type of well screen, covered with selected sand/gravel particles and bonded with resin adhesive.

borehole: a type of water well characterised by a relatively small diameter (100–450 mm) and large depth (usually more than 30 m).

borehole casing: steel or plastic pipe used to support the sides of a borehole. It may be slotted to allow water to enter and usually has screw joints.

borehole pump: see *submersible pump*.

break-pressure tank: an open tank positioned on a gravity pipeline to reduce the maximum pressure of water in the pipe.

cable tool: see percussive drilling.

carbon dioxide: a colourless gas which occurs naturally in the atmosphere. It is heavier than air and may accumulate in low-points giving rise to a non-breathable atmosphere.

catalyst: a substance which alters the rate of a chemical reaction but is itself unchanged by the reaction.

catchment: the area of land from which rain-water drains to a stream or river.

catchpit: a small chamber used to collect water from a spring or land drain.

cellulose: material obtained from wood pulp, cotton, etc., and used as a filter coating.

centrifugal pump: a pump which uses centrifugal force to move water.

cesspool: a brick- or concrete-lined tank used to hold sewage but not treat it. It needs periodic emptying.

chalk: a fine-grained white limestone found in east and south-east England. It is well known because of the "white cliffs of Dover" and forms an important aquifer.

chemical water treatment: a means of altering the chemical composition of water and disinfecting it by the addition of other chemicals.

chlorination: a method of disinfecting water using chlorine gas or a solution of sodium hypochlorite or bleaching powder, all of which contain chlorine.

cistern: a water tank which forms part of a water supply system.

climate: the variety of weather patterns which occur throughout the year over part of the earth's surface. The features of the weather patterns include such attributes as rainfall, dryness, temperature, windiness, etc.

common law: part of the British legal system which is based on precedence.

cone of depression: the inverted cone-shaped depression of the water table around a pumping well.

cone of exhaustion: an alternative term to cone of depression.

contract: a legal agreement between two parties which requires an offer by one side, an acceptance from the other and some "consideration" (i.e. payment or other benefit).

Cryptosporidium: a parasitic single-celled organism which uses animals and humans as a host; see *protozoa*.

derogation: when water is abstracted from a newly licensed source in such a way or at such a rate that people with a right to abstract water are prevented from obtaining a supply.

developing a well: increasing the yield of a newly constructed well or borehole by removing the sand, silt and mud which is adhering to the well face.

dew: see *precipitation*.

dew point: the temperature at which cooling moist air starts to deposit its vapour as fog or dew.

diatomaceous earth: a sedimentary rock made up of the remains of microscopic plants called "diatoms". It is exceedingly fine grained and absorbent and is used as a filter material.

discharge: the volume of water flowing in a stream or from a spring in a unit of time (e.g. litres per second). It is also the delivery rate of a pump.

disinfection: the reduction of the microbial contamination of water.

displacement pump: a type of pump which incorporates a piston to move water.

distribution system: the network of pipes, tanks, valves and taps used to transport water from a source or reservoir to the point where it is used.

divining: a mystical method of locating either groundwater or an auspicious place to construct a successful well using a hazel twig or other device. When submitted to impartial scientific tests divining has not been successful; also known as *dowsing* and *water witching*.

dolomite: a mineral composed of calcium and magnesium carbonate which is used to correct the acidity of water. It is also the name given to a type of limestone which is largely made up of dolomite.

dowser: see *divining*.

dowsing: see *divining*.

drawdown: the lowering of the water table in and around a pumping well. It is measured as the difference between pumping water level and the original or rest water level.

effective rainfall: total rainfall minus evaporation losses.

effluent seepage: water which escapes from the ground into a stream.

elevation: the height above sea level or another fixed level.

evaporate: see *evaporation*.

evaporation: the conversion of water into vapour.

evapo-transpiration: the combined water loss from evaporation and transpiration processes.

extraordinary use: the riparian use of water for industrial purposes; see *riparian use*.

falaj: an Arabian name for a horizontal well; see *qanat*.

filter: a device used to remove suspended matter from water as it flows through it.

fissures: natural cracks in rocks which greatly enhance groundwater movement.

flashy: the description applied to streams and rivers where flows peak soon after rainfall and then subside rapidly.

fog: minute particles of water suspended in the air; see *precipitation*.

foggaras: see *qanat*.

friction losses: the effect of water flow in a pipe; see *head losses*.

gauging station: a site selected on a stream or river where flows are recorded. It may incorporate a weir or other structure to improve the accuracy of the measurements.

Giardia: a parasitic single-celled organism which infects humans and animals; see *protozoa*.

granite: an igneous rock found on Dartmoor, the Lake District, Scotland and elsewhere.

gravel pack: gravel which is inserted around slotted screen in a borehole to prevent fine-grained aquifer material from being drawn into the borehole.

groundwater: water which is contained in saturated rock.

grout: a cement slurry used to fill the annular space between the ground and the lining of a well or borehole so that unwanted surface water cannot get into the well. It is important to use only the minimum quantity of water, otherwise the grout will shrink on setting and not make a seal.

hardness: the presence of certain dissolved substances in water which is characterised by difficulty in producing lather with soap. Temporary hardness is due to the presence of calcium and magnesium carbonate. It produces scale in pipes and kettles. Permanent hardness is caused by calcium and magnesium sulphates and does not produce scale.

head: the potential energy of water due to its height above a given level. It is used for the height to which pumps must lift water and also the pressure in a distribution system due to the presence of a reservoir. It is expressed in metres.

head losses: the pressure losses in an aquifer, well or pipe system expressed as a head of water.

humidity: a measure of water vapour in the atmosphere.

hydraulic gradient: the change of head within a water system such as pipework, stream or aquifer. Flow can only take place if a hydraulic gradient exists.

hydraulic ram: a device which uses the momentum of flowing water to force a small quantity of water into a delivery pipe.

hydrograph: the graph of river level or river flow plotted against time.

hydrological cycle: the series of interlinked processes which cause water to be circulated from the oceans to the atmosphere to the ground as precipitation and returned to the ocean as river flow.

impermeable: a description of rock, soil and other material which will not transmit water.

impounding works: the dam and associated overflow arrangements which have been made to hold water back in a reservoir.

influent seepage: the seepage of water out of a stream into the ground.

intake works: the structure and pipework used to take water from a stream or river.

interflow: that part of precipitation which percolates to a shallow depth before reaching streams and rivers. It can be identified as a separate part of a stream hydrograph.

ion: an electrically charged atom which forms part of a molecule.

ion exchange: a method of water softening in which calcium and magnesium ions are exchanged with sodium in a treatment resin.

irrigation: watering crops, usually by spraying water over them.

karez: see *qanat*.

laminar flow: a flow of water which is steady and continuous without any turbulence.

land drain: a system of porous pipes or culverts used to improve land by removing rain-water quickly and by lowering the water table.

land drainage: a system of keeping areas of land free of water by installing pipes, ditches, etc.

landfill: a site operated for the controlled disposal of waste materials by burial. The old-fashioned name is tip or dump, terms which now suggest a lack of control and consequential water pollution.

latitude: the extent to which a place lies either north or south of the equator.

levelling: a method of relating the height of one point to that of another.

limestone: a type of sedimentary rock composed largely of carbonate materials, particularly calcite and dolomite.

London Gazette: an official newspaper published weekly by the British Government and consists solely of legal notices.

meteorology: the scientific study of the atmosphere, especially with regard to weather and climate.

methaemoglobinemia: a medical condition in which nitrogen molecules have become trapped in the haemoglobin in red blood corpuscles and therefore prevent oxygen from being transported round the body.

minimum acceptable flow: the lowest flows which are needed in a stream or river so that there is sufficient water available for riparian abstracters, fisheries, navigation, sewage dilution and natural beauty.

nappe: a sheet of water flowing over a weir; it has an upper and a lower surface.

notch: the opening of a measuring weir; see *thin plate weirs, rectangular notch* and *V-notch.*

nuisance: a legal term which applies when one person's actions deprive another of his legal rights.

off stream reservoir: a reservoir which is located away from a stream, contrasting with an *impounding reservoir.* It is filled by pumping or diverting water from a stream.

ordinary use: riparian abstraction for domestic needs and stock watering; see *riparian use.*

outcrop: that part of a rock which is exposed at the surface.

peat: partially decomposed mass of vegetation which frequently has a high water content. This water is usually very acid.

perched water table: an area of saturated aquifer retained above the main water table by a small area of impermeable rock.

percolation: water moving downwards under the influence of gravity until it reaches the water table.

percussive drilling: a method of constructing a borehole using cutting tools which are suspended on a steel rope. The drilling action gives repeated blows to the rock at the bottom of the borehole. The broken rock is then removed using a bailer.

perforated casing: steel or plastic pipe which is used to line a borehole and has holes or slots cut in it to allow water to flow into the borehole.

permanent hardness: see *hardness*.

permeability: the capacity of soil and rock for transmitting water.

permit system: see *appropriation doctrine*.

pH value: a means of expressing the acidity or alkalinity of a solution. It is the logarithm of the reciprocal of the hydrogen ion concentration. The neutral point is pH value 7.0, with acids having lower values and alkalis having higher values.

pit latrine: a form of lavatory where the faeces accumulate in an underground pit.

plaintiff: the person bringing about a prosecution in a law suit.

pore spaces: microscopic spaces between individual grains in a rock. Good aquifers have interconnecting pores.

porosity: the percentage ratio of the void space to the total volume of rock.

precipitation: (1) water which falls from the atmosphere as rain, snow, sleet, hail, fog or dew; (2) the chemical process where material settles out of solution. This material is called the precipitate.

pressure tank: a storage tank for water which is kept pressurised in order to provide sufficient head to deliver the water in a distribution system. Pressure tanks are frequently used as part of small water supplies based on a borehole.

protected rights: in England and Wales, any abstraction which is made in accordance with an abstraction licence or is exempt from licensing takes precedence over any new licensed abstraction.

protozoa: a large group of single-celled animals which are different to bacteria and include a large number of parasitic forms which cause disease in humans and animals.

puddled clay: clay which has been made watertight by beating, rolling or other methods to reduce its permeability. Not all clays are suitable for this purpose.

pumping test: a method of testing a well or borehole to establish the reliable yield and to find out if a new well affects other wells and springs.

qanat: a sloping tunnel which connects a series of shafts or wells and intercepts both the water table and the ground surface some distance away, thus allowing groundwater to flow out at the surface. These structures are common in the Middle East and parts of Africa. They are known by a variety of local names including *karez*, *foggaras* and *falaj*.

rain-gauge: a device used to collect rain so that the quantity of rain can be measured. The rain is expressed as a depth equivalent and measured in units of length.

rainfall: see *precipitation*.

rain harvesting: a system of collecting rainfall from roofs, etc., in order to provide a water supply.

recharge: water which percolates to increase the quantity of groundwater stored in an aquifer.

rectangular notch: a measuring weir consisting of a thin plate of rectangular shape placed in a stream perpendicular to the flow.

reservoir: an artificial lake or a large tank used to store water under hygienic conditions. This last type of reservoir also provides sufficient head to enable water to flow through the distribution system.

rest water level: the water level in a well or borehole which is unaffected by pumping. It represents the local level of the water table.

riparian doctrine: the legal right to abstract water which applies in the eastern part of the USA. It is also called the rule of reasonable sharing or the rule of correlative rights and is similar to the system of riparian use in the UK; see *riparian use*.

riparian occupier: one who occupies land which adjoins a water course and as such enjoys rights over the water.

riparian use: in the UK this may be ordinary use which is generally the reasonable use of water for domestic purposes and watering stock, or extraordinary use where the water is used for industrial purposes. In both cases, the riparian user must take account of the impact on other riparian users.

rotary drilling: a rapid method of drilling boreholes. It uses cutting tools suspended on screw-jointed rods which are rotated to produce the cutting action. The broken rock is removed by the circulation of a fluid in the borehole. Fluids used include air, water, special drilling foam and various types of mud.

rotary pump: a pump which has rotating elements and is capable of raising large volumes of water at low pressures; see *rotor* and *stator*.

rotor: a length of steel rod shaped into a slow spiral used in rotary pumps to move water when it is rotated inside a stator; see *stator*.

rule of absolute ownership: a system of water rights in the USA which allows unlimited use of groundwater from beneath an owner's land.

rule of reasonable sharing: see *riparian doctrine*.

run-off: that part of precipitation which flows across the ground surface into streams. It forms that part of stream flow which most rapidly decreases once the rain has stopped.

salination: the build up of mineral salts in the soil caused by the evaporation of irrigation waters. The process may also lead to increases in the concentration of these minerals in the groundwater as a proportion of the surface deposits are dissolved by percolating irrigation water.

salinity: the total concentration of dissolved minerals in water.

sand trap: a chamber or tank placed in a pipeline to allow sand and silt to settle out.

septic tank: a tank in which sewage is treated by bacterial action.

silage: an animal fodder made of finely cut green plants packed tightly in tanks (silos) and fermented. A highly polluting liquor is produced.

siliceous rock: rocks which are made of silica such as most sandstones and volcanic rocks, as opposed to limestone which is made of calcite. Siliceous rock are used for gravel packs around wells and for soakaway systems because they are not dissolved by percolating waters.

site investigation: an investigation to find out whether geological conditions are suitable for the foundations of a dam or earth embankment.

slotted casing: borehole lining of steel or plastic which has saw cuts to permit the entry of water.

snifter valve: a valve designed to admit a small quantity of air into an air vessel on a hydraulic ram. This keeps the vessel fully charged.

soft water: relatively pure water which lathers easily when mixed with soap; see *hardness*.

source: a place from which a supply of water may be obtained.

source works: a construction which makes it possible to obtain a supply of water and from which the supply is taken.

specific capacity: the rate at which a well or borehole can be pumped for a unit of drawdown. It is often shown in graphical form, the specific capacity curve, and is used to check the efficiency of the well.

spillway: part of a dam which is designed to allow water to flow over it without endangering the stability of the dam.

spring: a natural outflow of groundwater at the ground surface.

stage discharge relationship: the relationship between river level and flow which enables a continuous record of flows to be made by monitoring levels.

stator: a flexible moulded sheath inside which the rotor of a rotary pump is turned to produce a pumping action.

statute: a law passed by the government of a country.

statute law: that part of the British legal system which is made up of Acts of Parliament.

steening: the brick or stone lining to a hand-dug well.

storage coefficient: the volume of water held in an aquifer which can be released by gravity drainage. It is expressed as a percentage of the total volume of the aquifer.

strainer: a perforated pipe, cylinder of wire gauze or other similar device which is fitted to the end of a pipe or pump inlet to exclude large particles.

submersible pump: a pump which is designed to operate under water. Usually, these are electrical centrifugal pumps and have the electric motor enclosed in a waterproof casing. Sometimes these pumps are driven by rods which are rotated by a surface motor.

surface run-off: the part of rainfall which flows over the surface to streams; also the element of stream flow so formed.

surface water: water which is contained in rivers, streams, lakes and ponds. Springs represent the point where groundwater becomes surface water.

swabbing: moving a plunger up and down a borehole to force water rapidly in and out of the borehole wall. It is a method frequently used for borehole development.

temporary hardness: see *hardness*.

thin plate weirs: a weir consisting of a thin plate with a rectangular or V-shaped notch cut into it and set at right angles to the flow of water.

tile drains: usually unglazed earthenware pipes, but sometimes concrete pipes, laid with open joints to receive and remove excess water from the soil.

transpiration: the evaporation of water from the leaves of plants. Water is drawn up through the plant structure, which in turn sucks water into the plant from the soil. This mechanism accounts for the majority of water returned to the atmosphere on land.

trial pit: a small pit dug to provide soil samples as part of a site investigation.

turbine pump: a type of rotary pump with the water entering at the centre and moved by rotating vanes.

turbulent flow: when water flows in an unsteady state and with a great deal of disturbance.

ultraviolet radiation: a form of light which can cause sun-burn, etc., and is produced artificially using mercury vapour lamps. It is used as a disinfectant for water as it has strong action in killing microbes.

underground strata: a legal term for all geological materials below the ground surface.

underground water: an alternative term for groundwater frequently used in a legal context.

V-notch: a type of thin plate weir where a notch of 90°, ½ 90° or ¼ 90° is cut into the plate.

vacuum: a space which contains no molecules. A perfect vacuum is virtually unobtainable.

valve: a mechanical device fitted to a pipe to control the flow of water. A tap or faucet is the best known type of valve.

vapour: a gas from a substance which is a solid or liquid at normal temperatures.

velocity–area method: a method of calculating the flow of water in a stream channel by multiplying the measured average velocity of the water with the cross-sectional area.

viruses: disease-producing microscopic particles which can only reproduce within a living cell.

water in underground strata: a legal term for groundwater.

water table: the upper surface of the saturated part of an aquifer.

water witching: a term for *divining* or *dowsing*, used in North America.

weather: the local state of the atmosphere at a particular time expressed in terms of temperature, amount of cloud, precipitation, sunshine, etc.

Weil's disease: a form of jaundice also known as leptospirosis.

weir: a structure used to control the flow of water, often so that it can be measured.

weir tank: a device used to measure the flow of water discharging from a pump. It consists of a tank which incorporates a weir.

well: a pit, excavation or borehole constructed to obtain a water supply from groundwater.

well face: the inner wall of a well or borehole through which water enters.

well screen: a type of borehole casing specially designed to allow the efficient entry of water into the borehole.

wind pump: a form of piston or rotary pump which is powered by a vane rotated by the wind.

yield test: a pumping test designed to determine the safe yield of a well or borehole. The test consists of pumping at high rates for a sustained period while monitoring the discharge rate and water levels in the pumping well, surrounding wells and local spring flows, etc.

Index

Abstraction licence, 60–2, 202, 204–6, 209–12
Acid rain, 242
Agricultural abstraction
  water requirements, 66
  water rights, 198, 200–1
Air-borne pollution, 124, 242
Air-lift pumping, 42–6
Algal growth, 64, 73
Anti-seepage collars, 89, 229
Appropriation doctrine, 213, 215–18
Aquifer, 14–20, 22, 56, 76, 80–1, 83, 105, 108, 110–11, 114, 211, 224, 227, 229, 238–43
  confined, 15, 16
  tests, 60–1
Artesian
  aquifer, 16–17
  flow, 16, 139
  wells, 16–17
Atacama Desert (Chile), 76–7
Auger, 97–8, 157

Bacteria
  in sewage treatment, 135
  in tanks, 76, 186

in water supply, 115, 120, 124, 136, 185
in wells, 56
treatment for, 125–7, 131, 135, 194
Bleach, 81, 127
Boreholes
  casing, 18, 45, 112, 211, 240–1
  cleaning, 187–9, 224–5,
  design and construction, 95–114
  efficiency, 187–190
  exploratory, 100
  gravel pack, 110–11, 114
  headworks, 109–11, 187
  maintenance, 187–9
  protection from pollution, 109–11
  screen, 113–14
  source of water, 25
  yield development, 106–8
Bottled water, 69, 136

Cable tool drilling, 102–4
Catchpit, 79–82, 128, 187, 232, 242
Cess pool, 234–6
Chile, 76–7
Chlorination, 126–9
Climate, 8

Collection chamber
  groundwater, 82–3
  spring, 79–82, 187, 232
Cone of depression, 17–20
Consultants, 4–5, 205, 212
Contracts, 101
Conversion factors, 224–5
Costs
  boreholes, 197
  energy, 24
  leakage, 69–70
  NRA, 208–9
  pumping, 16, 18, 57, 85, 145, 164, 224
  reservoirs, 84–6
  treatment, 186
Cryptosporidium, 125

Denmark, 3, 218
Derogation, 201, 205, 212, 222, 232
  effects of, 223–4
Desalination, 73
Dew, 8–10, 76
  point, 10
  pond, 76
Discharge
  from an open pipe, 47–9
  from pumping, 45, 238
Disinfection, 127–9
  pipes and tanks, 81, 185
  rain-water collection tanks, 74
  wells, 95
Ditch, 37, 45, 81, 109, 175–7, 182, 230–1
  pipe crossing, 175–7
  pollution from, 232–3
Domestic water supplies
  quality, 116–18, 136–7
  quantity, 63–5
Drawdown, 18–20, 42, 57–9, 148, 189
Drilling, 2, 90, 95–110
  cable tool, 103–4, 107–8

contractor, 99–101, 188–9
  contracts, 101
  down-hole-hammer, 106
  fluids, 106
  percussive, 103–4, 107–8
  rotary, 105–7
  tube well, 95–9
Drought, 73, 223
Dysentery, 124

EEC Directive, 136–7
  mineral water (89/397/EEC), 116–18
  water for human consumption (80/778/EEC), 116–18, 136–7
Embankments, 86–90
Engineering works, 224–9
England, 2, 4, 10, 73, 76, 79, 107, 124–5, 137, 140, 148, 175, 197, 199, 200, 242
Environmental health, 22, 115, 137, 196
Estate water supplies, 196
Evaporation, 7, 10–11, 138
  losses, 31, 64–5, 83
Evapo-transpiration, 11
Expert
  employing one, 4–5

Farm water supplies, 1, 25, 101
  animals, 63, 66, 68
  crops, 24, 66–7
  water quality, 119, 125, 137–8
Filters, 129–31, 135
  activated carbon, 132, 135
  porous pot, 130
Fire-fighting, 65, 169–70, 201
Fish farming
  sources, 68, 83
  water quality, 137–8
  water requirements, 68

Flow measurement, 31–41
  jug and stop-watch, 39–41
  stream flow, 38
Fog, 8–10, 76
  harvesting, 76–7
France, 16, 219, 235
Friction losses in pipes, 148, 166–8,
    171, 179, 182
Frost
  protection from, 67, 83, 175–8,
    182, 190–1, 193, 206

Geological considerations, 13–14,
    22, 86–7
Germany, 3, 218–19, 242
Giardia, 125
Glass-houses, 67, 75–6
Gravel pack, 94, 96, 111, 113–14
Gravity flow, 24, 65, 68, 80, 139–41,
    167, 177, 183
Groundwater, 7, 14–20
  abstraction permits, 119, 201,
    210–12, 213, 216–17, 219,
    220–1
  collection systems, 82–3
  flow, 17
  riparian rights, 198, 213–15

Hardness, 120–1, 131–3
Hydraulic ram, 147–53, 191
Hydrograph, 12, 13, 39
Hydrological cycle, 6–20

Interflow, 12–14
Ireland, 2, 22, 197, 200
Iron removal from water, 134–7
Irrigation, 24, 66–7
  glass-houses, 75–6
  groundwater lagoons, 82–3
  riparian rights, 197–8

water quality, 119, 137–8
water requirements, 67
Isles of Scilly, 73
Italy, 3, 219

Japan, 8, 220

Laboratory analysis, 110, 115, 120
Lagoon, 82–3
Land drainage
  as a water supply, 22–3
  construction, 80–2
  effect on supplies, 229–31
Landfill, 118, 242–3
Latrine, 236–7
Lawyer, 5, 198, 223, 229
Leak detection, 70, 192–3, 237–8
Legionella, 124
Levelling, 141–5
Licences
  abstraction, 4, 60–2
  application for, 60–1, 201–5,
    206
  charges, 208–9
  environmental impact, 205
  impounding, 79, 209–10
  protection given by, 208
  pumping tests, 60–1

Mains water supply, 1–2, 64, 73,
    126, 241
Maintenance, 186–94
  co-operation, 23, 80
  need for, 70, 83, 115, 125, 186
  pipes, 192–3
  pumps, 191–2
  spring sources, 187
  storage tanks, 190–1
  treatment systems, 125, 194
  wells and boreholes, 187–90

Methaemoglobinemia, 240
Mining
    impact on water supplies,
        241–2

National Rivers Authority, 200–12
Netherlands, 220
Nitrate
    in water, 120
    pollution from, 239–41

Percussion drilling, 103–4, 107–8
Pesticides, 123, 239–41
pH, 116–18, 121–3
    treatment, 123, 134
Phenol, 127
Pipes, 165–77
    capacity, 167, 171, 173–5
    choosing, 165–7
    disinfection, 127–8
    flow from, 47–9
    flow in, 167–8
    friction losses, 148, 166–8, 171,
        179, 182
    layout, 168–71
    leak detection, 192–3
    maintenance, 192–3
    materials, 171–2
    protection, 175–7
    safety, 45
    size, 171–2, 173–5
Pollution
    air-borne, 242
    boreholes, 109–11
    land drains, 80
    landfill, 242–3
    nitrate, 239–41
    pesticide, 239–41
    road run-off, 243
    rubbish tips, 242–3
    silage, 239

spillages, 238–9
springs, 79–81
streams, 22, 77
wells, 94–6
Ponds, 12, 64–65, 68, 83, 115, 137,
    169–170
dew, 76
Precipitation
    chemical, 56, 134–5
    hydrological cycle, 8–10, 29
Private supplies
    number of, 3–4
Protected Rights, 201
Protozoa, 124–5
Pumps, 144–65, 225
    air-lift, 42–5, 188–9
    centrifugal, 159, 160–2
    choosing, 41–2
    constant displacement, 154–60
    deep well, 158–60
    helical rotor, 160–1
    how a pump works, 145–8
    hydraulic ram, 147–53
    jack, 157
    jet, 163–4
    maintenance, 191–2, 231
    reciprocating, 154–9
    rotary, 159–61
    submersible, 160–2
    suction, 146–8
    variable displacement, 159–64
    water turbine, 154
    wind powered, 164–5
    wooden, 156–7
Pumping tests, 41–61
    aquifer, 60–1
    consent, 210–12
    constant rate, 60–1
    licensing, 60–1
    planning, 55–61
    step, 58, 189–90, 232
    water level readings for, 55–7
    yield, 58–60, 189–90

Qanats, 24–5
Quarries, 224–6, 241–2

Rainfall, 8–9, 12, 21–2
  average annual, 9
  measurement, 27–31
  source of water, 21–2, 72–6
  statistics, 9
Rain-gauge, 27–30
Rain-water collection, 21–2, 30–31,
  72–6
Reservoirs, 22, 83–90
  construction, 85–90
  impounding, 84–90
  off-stream, 84–5
  storage, 181–3, 190–1
Reverse osmosis, 135–6
Riparian doctrine, 213–4
Riparian rights, 197–8, 201, 218
Rivers
  flow measurement, 31–9
  flow records, 31
  hydrological cycle, 7, 12–14
  intakes, 77–9
  source of water, 22
Roads
  cuttings, 227, 229
  run-off, 243
Roof drainage, 30–31, 72–6
Rotary drilling, 105–7
Rubbish tips, 188, 242–3

Safety, 246–50
  air-lift pumping, 45
  chlorine solution, 127–8
  tanks, 191
  ultra violet light, 128–9
  wells, 91, 93, 186–8
Salination, 138
Sampling from well, 54–5
Sand trap, 184–5

Scotland, 2, 7, 137, 197, 200
Septic tank, 77, 95, 110, 187, 222, 232,
  234–6
Sewage, 70, 120–1, 124, 187, 205, 232
  treatment, 233–7
Sewers, 95, 110, 222, 232
Silage, 239
Silt
  cleaning from wells and
    boreholes, 107–10, 187–9
  trap, 73–4
  weirs, 35, 37
Singapore, 220
Sodium in water softening, 131–3
Sodium hypochlorite, 95, 127
Spain, 3, 24–5, 220
Specific capacity, 57–9, 189–90
Spillage
  pollution from, 238–9
Spillways, 85–6, 88–9
Springs
  borehole pumping, 19–20
  collection chamber, 79–82
  flow measurement, 39–41
  hydrological cycle, 12–16
  land drainage impact of,
    228–9
  quarries impact of, 222
  road drains impact of, 227, 229
  source of water, 22–5
Storage tanks
  leaks from, 237–8
Streams
  borehole pumping, 19–20
  flow measurement, 31–9
  hydrological cycle, 12–16
  intakes, 77–9
  source of water, 22
Support to land
  loss of, 199
Surface run-off, 12–14
Surveying, 141–5
Swimming pools, 65

Tanks, 177–81
  break-pressure, 179, 183–4
  dead animals, 179
  frost protection, 178–9
  maintenance, 190–2
  plastic and fibreglass, 179
  pressure, 179–81
  rain-water, 73–4
  selection, 177–9
  settlement, 184–5
  weir, 45–6
Taste, 121, 127, 135–6, 179
Technical terms, 5, 253–63
Thin plate weirs, 31–7
  installation, 33–5
Transpiration, 5

United Kingdom, 2–3, 9, 195, 197,
  212
United States of America, 3, 8–9, 17,
  23, 112, 137, 235, 237–8
  abstraction laws, 195–7, 212–18
  drinking water standards, 116–18
Units, 224–5

Valves
  controlling pumping rate, 58
  in supply system, 139–40, 172,
    175, 192–3
Vermin
  protection from, 73, 81, 109, 179,
    182–3, 185, 190–1
Viruses, 124, 130

Wales, 2, 4, 79, 137, 197, 199–200
Water divining, 25–27
Water level
  measurements, 48–55
  measuring device, 51–55

pumping, 18–19
pumping test, 57
rest, 18
Water meter, 45, 191–2
Water permits, USA, 215–7
Water quality
  acid water, 121–3, 134
  agricultural requirements, 137–8
  characteristics, 120–5
  deterioration, 232–43
  guidelines, 116–18, 136–7
  hardness, 120–1, 131–3
  legal controls, 136–7
  treatment, 125–36
Water requirements
  agriculture, 63, 67
  domestic, 63–5
  fish farming, 68
  irrigation, 66–7
  small businesses, 63, 68–9
  trees, 11
Water Resources Act , 200–12
Water rights, 195–212
  appropriation doctrine, 213,
    215–8
  Denmark, 218
  estates supplies, 196–7
  France, 219
  Germany, 218–19
  Italy, 219
  Japan, 220
  Netherlands, 220
  riparian doctrine, 213–14
  riparian rights, 197–8, 201, 218
  Singapore, 220
  Spain, 220
  UK, 197–212
  USA, 212–18
  Water Resources Act 1991, 200–12
  Zimbabwe, 221
Water savings, 69–71
Water table, 14–16
  fluctuations, 19–20

lowering by drainage schemes,
    229–32
lowering by pumping, 18, 19
lowering by quarrying, 223,
    226
perched, 15
Water treatment, 125–36
    activated carbon, 132, 135
    bacterial, 125–7
    desalination, 73
    disinfection, 127–8
    filters, 129–31, 132, 135
    iron and manganese removal,
        134–5
    pH adjustment, 134
    reverse osmosis, 135–6
    softening, 131–3
    ultra violet radiation, 128–9
Water troughs, 23, 80, 169, 170
Water witching, 25–7
Weil's disease, 179
Weirs, 31–7
    choosing, 37
    rectangular, 36
    tank, 45–6
    thin plate, 31–7
    v-notch, 32–3, 35–6
Wells
    artesian, 16–17

construction, 90–6
disinfection, 95
face, 17–18
festivals, 26
flow into, 17–20
gas, 188
maintenance, 187–9
safety, 91, 93, 187–9
screen, 133–4
siting, 25–7
source of water, 25
testing, 41–61
tube well, 95–9
Wind pumps, 164–5
World Health Organisation, 116–8,
    127, 136

Yield
    development of, 106–10
    estimate of, 11, 30–1
    loss of, 189–90, 222–32
    reliable, 21
    safe, 55
    tests, 58–60
    wells and boreholes, 18, 41–61

Zimbabwe, 221